EMERGENCY RESPONSE

LIFE, DEATH AND HELICOPTERS

DAVE GREENBERG
with Adrienne Kohler

RANDOM HOUSE
NEW ZEALAND

RANDOM HOUSE

UK | USA | Canada | Ireland | Australia
India | New Zealand | South Africa | China

Random House is an imprint of the Penguin Random House group of companies, whose addresses can be found at global.penguinrandomhouse.com.

First published by Penguin Random House New Zealand, 2017

10 9 8 7 6 5 4 3 2 1

Text © Dave Greenberg, 2017

The moral right of the author has been asserted.

All rights reserved. Without limiting the rights under copyright reserved above, no part of this publication may be reproduced, stored in or introduced into a retrieval system, or transmitted, in any form or by any means (electronic, mechanical, photocopying, recording or otherwise), without the prior written permission of both the copyright owner and the above publisher of this book.

Cover and text design by Rachel Clark © Penguin Random House New Zealand
Cover photograph by Scott Hammond/*Marlborough Express*
Back cover photograph by Andrew Gorrie/*Dominion Post*
All other photos and illustrations, unless otherwise
credited, are courtesy of Dave Greenberg
Prepress by Image Centre Group
Printed and bound in Australia by Griffin Press, an Accredited
ISO AS/NZS 14001 Environmental Management Systems Printer

A catalogue record for this book is available from
the National Library of New Zealand.

ISBN 978-0-14-377166-1
eISBN 978-0-14-377167-8

penguin.co.nz

EMERGENCY RESPONSE

DEDICATION

THERE ARE MANY people who have helped me to create a life better than I could have ever imagined and to complete this book.

Firstly, my parents, Judy and Bill Greenberg, who taught me right from wrong, showed me what good manners were and why they mattered, always wanted me to do well, put up with a lot of crap from me in my teenage years, and stuck by me across all the miles. You didn't always agree with or like my decisions (especially staying in New Zealand!) but you never stopped encouraging or loving me.

Foster Schucker who mentored my IT career and was instrumental in my move to New Zealand.

My New Zealand family and friends who have shared, and seen me through, the highs and lows over the past 25 years. Your phone calls, texts, meals and hugs have meant so much to me. You are the reason that I have coped with the things I have seen, done and been part of.

Georgia and Zara. I have known both of you since you were hours old, and you have kept me amazed and amused ever since. Watching you grow up, and seeing life through your eyes, has been life-changing. You are proof that it is easier, cheaper and more fun to be an 'uncle' than a dad.

John Goldswain, who brought me into the Life Flight family in 1991; although we didn't always see eye to eye, you were an amazing boss, teammate and friend. You were taken way too early my friend, but you made an incredible, positive, difference to my life and thousands of others.

Adrienne Kohler. Who would have thought a chance conversation at a pub quiz would end up in you becoming my editor? Thank you for helping me lift my story from a police report to the book it has become.

The many friends and colleagues who pre-bought the book, believing in me and my ability to get it done.

Harry Mills and the others at Bizdojo, Wellington, who constantly motivated me with ideas, encouragement, coffee and whatever else it took to get the book done!

Lastly, to all the incredible pilots, crew, paramedics, police, firefighters, doctors, nurses, dispatchers, call-takers, flight coordinators, engineers, administration and fundraising people I have worked with. Saving lives is a team effort and I have been privileged to work with some of the most compassionate, dedicated and skilled people in the world.

PREFACE

I HAVE COMPILED this book using my memories, logs, media reports and by speaking with some of those involved. The least reliable of all these is my memory! Over time waves seem to get bigger and winds stronger.

With nearly 4000 missions to recall, I have only managed to look back on a few. I hope that I do not offend any of my workmates or colleagues if you are not specifically mentioned.

Many people were uncomfortable with their surnames being used, so I made a blanket rule that unless I have specific permission, I have only used someone's first name. Some people have asked not to be identified at all.

Any patient identified in the story has given permission for their story to be told through my eyes. I am very grateful that you have trusted me to tell it and I hope I have told it well.

In most cases I have used UK spelling, as opposed to USA spelling. I have converted all measurements to the metric system, which will be particularly strange for the aviators reading this.

TABLE OF CONTENTS

13 PROLOGUE

15 ONE
Growing up in New York

21 TWO
Hating School

30 THREE
Early Ambulance Missions

38 FOUR
University

44 FIVE
Dallas

49 SIX
First Trip to Wellington

56 SEVEN
Crossroads

58 EIGHT
The Move to New Zealand

63 NINE
Money and You

67 TEN
Welcome to Life Flight

73 ELEVEN
Dope on the Rope

81 TWELVE
Learning to be a Winch Operator

87	THIRTEEN Media Spokesperson
94	FOURTEEN Teamwork and Crew Resource Management
102	FIFTEEN Search and Rescue
106	SIXTEEN No Room to Rescue Them All
110	SEVENTEEN Hospital Transfers — ICU
114	EIGHTEEN Hospital Transfers — NICU
117	NINETEEN Sounds Air Crash
124	TWENTY Helmet and Winch Cam
127	TWENTY-ONE Terminator
140	TWENTY-TWO Responding to the Scene
143	TWENTY-THREE September 11, 2001 — NZ Time
150	TWENTY-FOUR September 11, 2001 — US Time

154 **TWENTY-FIVE**
Road Traffic Crashes

159 **TWENTY-SIX**
Into the Night

162 **TWENTY-SEVEN**
Our Helicopter Crash and New Air Rescue Centre

166 **TWENTY-EIGHT**
Rural People are Tougher than City Folk!

170 **TWENTY-NINE**
Imagine It. Done. — Part 1

179 **THIRTY**
Imagine It. Done. — Part 2

187 **THIRTY-ONE**
Rescue Dave

193 **THIRTY-TWO**
Anzac Day — Part 1: Conflicting Missions

198 **THIRTY-THREE**
Anzac Day — Part 2: The Search and Rescue

205 **THIRTY-FOUR**
Anzac Day — Part 3: The Aftermath

214 **THIRTY-FIVE**
Suicide and PTSD

218 **THIRTY-SIX**
The Special People in My Life

223 **THIRTY-SEVEN**
Christchurch Earthquake

228 **THIRTY-EIGHT**
Police Missions

240 **THIRTY-NINE**
My Longest and Most Difficult Flight

248 **FORTY**
When Things Go Wrong — A Few Close Calls

259 **FORTY-ONE**
Management Team

265 **FORTY-TWO**
Management Review — The Beginning of the End

269 **FORTY-THREE**
End of My Flying

276 **FORTY-FOUR**
Changes Over Time

279 **FORTY-FIVE**
New Beginnings

PROLOGUE

IT'S EVERY RESCUE team's worst nightmare and, like every nightmare, it seemed to happen in slow motion. I was standing in the doorway of the rescue helicopter; gale-force winds pounding us above a churning, angry sea. We were winching nine sailors off a stricken yacht — two were already safely in the helicopter, and the next two were on the winch hook and approaching the skid. So far, the rescue had gone as smoothly as could be expected in such terrible conditions. The two sailors were only a metre below the skid, seconds from safety.

As we watched, one of the sailor's bodies went limp; he let go of the hook, his arms came up, and he slipped out of the rescue harness, plunging nearly 10 metres into the turbulent waves, disappearing below the water.

We were running low on fuel and only had minutes to find him — if he was even on the surface. The 40-foot yacht we had rescued him from had been difficult to spot in these seas, so we knew the chances of spotting him were slim.

Minutes later, I was on a winch line being dragged through some of the biggest waves I had ever been in, desperately searching for him. As I came through a breaking wave, there he was, directly in front of me. He reached out for me as I reached for him. I wasn't sure which of us was happier to see the other. I shouted at him over the howl of the gale and hovering helicopter: 'G'day, I'm Dave.'

ONE
GROWING UP IN NEW YORK

I WAS 13 years old the first time I kissed another person on the lips. It was in the middle of a New York City sidewalk, and it was nothing like the soft tender kiss that my friends at school bragged about.

The guy was a stranger to me; I didn't know his name, I knew nothing about him except he had a two-day stubble on his cheeks and smoker's breath.

My first kiss was the kiss of life, and it was the best first kiss I could have ever wished for.

It was a few days before my fourteenth birthday in August 1975, and I was on my lunch break from my summer job working for a family friend in Manhattan.

I hated being in the city on hot, humid days. As soon as I walked out of the cool, air-conditioned shop, the heat would hit, and I'd instantly break out in a sweat. The smell of diesel and petrol fumes mixed with rotting garbage would hit me like a brick wall, as did the noise of horns and sirens, the sound amplified as it bounced off the buildings that line New York's canyon-like streets. As always, cars, buses, trucks and yellow cabs hustled each other to navigate the never-ending gridlock, while locals impatiently skirted around the hordes of camera-toting tourists craning their necks at the sky scrapers above.

Just another day in New York! As I walked along, I was dreaming away as teenagers do and trying to make my biggest decision of the day — what to have for lunch — right until the man collapsed on the sidewalk metres in front of me.

I have to confess my first response was excitement — finally, I had my chance to use some of my training!

As I ran towards him, I mentally ran through the drills I had been taught. Safety was the top priority, so after a quick check to make sure the area was free of danger, I knelt down next to him and shook him while I asked if he was okay. He didn't respond. I rolled him onto his back, tilted his head back and looked, listened and felt for signs of breathing. My excitement turned to fear when I realised he was not breathing.

But I didn't hesitate. Back then it was still considered okay to give mouth-to-mouth rescue breaths to a stranger without a protective mask, so I gave him those first kisses of life — two big rescue breaths. I checked his carotid artery for a pulse — nothing. He was in cardiac arrest.

Then my training kicked in — I had prepared for this day. Even though I was only a teenager, I took charge and people listened. I directed one of the passers-by, who always seem to gather, to go to the payphone on the corner and call 911. I instructed another one to keep people back, and then I called out to see if anyone else knew CPR, but no one came forward.

Unfortunately, it was all an act. Although I seemed calm, cool and collected, I was terrified. The man was dead, and if I didn't do a good job, he would stay dead. His life was in the hands of a 13-year-old child!

I loosened his tie and ripped his shirt open to expose his chest. I quickly ran through the process I had practised on a dummy so many times — located the bottom of his ribcage, followed it to where it joined the other side then placed the palm of one hand two finger lengths above that, on his sternum, where I had to press down to do chest compressions.

I was about to start when — to my great relief — a woman came up who said she also knew CPR. Relieved to share the responsibility of trying to save this guy's life, I began chest compressions while my new-found lifesaving partner got ready to give him rescue breaths.

I pushed down hard and steady. The cadence of two-person CPR is easy. As I pushed down, I counted out loud — 'One-one thousand, two-one thousand, three-one thousand, four-one thousand, five-one thousand, breathe' — and when I said 'breathe' the woman made sure his head was tilted back, gave a large breath, and we watched his chest rise. As his lungs expelled the breath, I began chest compressions again: 'One-one thousand, two-one thousand . . .'

As we did this, a few people gathered to watch, but most rushed by barely glancing and probably more worried that they were being delayed by the crowd on the sidewalk than about a dying man.

CPR is incredibly hard work, particularly if you are doing it on a hot, humid New York afternoon. After a few minutes, the woman and I switched places. Here we were, two strangers performing CPR on another stranger on the sidewalk, in the middle of New York City. It was surreal.

Sometime later, I'm not sure if it was 2, 10 or 20 minutes, we heard the sirens of an ambulance approaching. Two medics appeared at my side with equipment in their hands. One of them checked the man while the other set up their equipment. As they connected heart monitor leads to his chest, they told us to stop CPR. We anxiously stared at the heart monitor, but all it showed was a lifeless rhythm. The medic asked us to continue CPR, I restarted chest compressions and the woman rescue breaths, although she now had a demand valve, which delivered 100 per cent oxygen whenever a button on the mask was pressed.

Soon they were ready to use the defibrillator, a machine with two paddles that delivers an electrical shock to the heart. Although I had seen one at the ambulance station and during my CPR training, I had never witnessed one being used, except on TV.

While one medic rubbed the two chest paddles together, spreading a gel across them, the other pressed a button as they waited for the machine to generate enough voltage. Suddenly the medic with the paddles called out: 'Stop CPR and stand clear!' As we leaned away, he glanced around to make sure we were indeed clear, placed the paddles on the man's chest and pressed a button.

Zap! I was startled when the man's limp body jumped slightly as the electric shock was delivered. The medics looked at their monitor and then at each other — one shook his head, and the other shouted 'clear' again and another shock was delivered. Again, the man's body jumped, and I could smell burning as the electric paddle scorched his skin.

Suddenly the heart monitor burst into life. The lifeless rhythm was replaced by a rhythm that I had seen on *Emergency!*, my favourite TV show, countless times before. He was alive!

The man's heart was pumping, and he was breathing on his own. The woman and I were ecstatic, although the medics looked like they had seen it all before. I didn't care; my heart burst with pride when they told us that we had saved his life.

We helped load the stretcher in the back of the ambulance, and the driver shook my hand and congratulated me again as he closed the back door. The woman and I hugged as the ambulance drove away with the lights flashing and siren blaring. I never did learn her name or the name of the man who we had saved. Forty-two years on, I don't have any idea if he died later that day or if he is still alive and kicking today.

None of that matters. All I knew for sure is that I had helped save someone's life. My childhood dream of becoming a firefighter and paramedic had just become more real.

I SAY I AM a native New Yorker because I grew up in part of New York City. And, being a real New Yorker, I know that 'the city' is Manhattan. When people dream of visiting New York, they are dreaming of visiting Manhattan, the New York they see on TV and in the movies. But my neighbourhood was just like any other suburb in every other big city in the world — and no one dreams of visiting the suburbs.

I grew up in Howard Beach, Queens, one of the many neighbourhoods that make up New York City. In the 1960s and '70s it mainly comprised white, middle-class Jewish and Italian families. I was part of a Jewish family of five: my mum Judy, my dad Bill, my older sister Linda and my younger sister Arlene.

We were a close family, and I grew up knowing what it meant to be loved and cared for. My parents worked hard to make sure that we had everything we needed and a lot of what we wanted. It was the era when dads would go out to work while mums would stay home and look after the house and family. We were no different.

My friends and I played outside until dark in good weather or gathered at one house or another in bad weather. If my mum was looking for me she would yell out the front window. By the tone of her voice, I could tell if I was being called in for dinner or if I was in some sort of trouble. If I didn't respond then a series of phone calls would take place, starting with the house I told her I would be at, and if I had moved on, continuing until she tracked me down.

Just to the east of Howard Beach, only a couple of kilometres away, sits JFK Airport, one of the busiest in the world. My house sat under one of its many flight paths, with planes landing or taking off constantly, only a minute or two apart during busy periods. Conversations had to briefly stop, or voices had to be raised to be heard above the sound of the engines as the planes passed overhead. The aircraft passed so low that I could just about make out the people sitting by the windows. I sat in my backyard waving up at them, and I was sure they waved back at me. I spent a lot of time imagining where they were heading or where they had been — it was a big world out there, and the passing planes made me want to explore it.

From the western edge of my neighbourhood, just a few blocks from home, I could see the Manhattan skyline, less than 15 kilometres away. My life was far removed from the glitz and glamour of Broadway, and the wealth and power of Wall Street, but there it stood, so close, yet so far. I remember watching the Twin Towers rising into the skyline, growing at the same time as I did. The Towers, we were told, were part of what made New York and America great.

I believed it. Decades later, as I sat staring at my TV watching those towers crumble to the ground, I felt like a huge part of my childhood crumbled with them.

Perhaps the oddest thing about unremarkable, average Howard Beach was the many Italian Mafia families who also lived there. The 'Dapper Don' himself, John Gotti, who went on to be the boss of the infamous Gambino crime family, lived just a few blocks from my home.

The 'Mafia houses' looked much the same as any of the other houses in the neighbourhood, apart from the undercover FBI cars that regularly parked outside them. In a suburb full of family wagons, these nondescript cars were the easiest way for us to figure out which houses held the prominent Mafia family members because at one time or another they would end up with the FBI parked outside.

Luckily, none of these guys seemed to bring their work home with them. Oddly, I think Howard Beach was made safer thanks to all our Mafia neighbours. Maybe burglars wanted to avoid accidentally burgling a Mafia member's home, or perhaps it was the added security of FBI cars continually staking out the neighbourhood?

The 2017 view of the NYC skyline from the edge of Dave's neighbourhood.

TWO
HATING SCHOOL

AS PART OF New York City, our emergency services were provided by the New York Police Department, NYC EMS (Emergency Medical Service) and the Fire Department of New York — the brave men (no women back in the '70s) of the FDNY.

Like many parts of New York, we had a volunteer ambulance corps which responded to emergencies in parts of the neighbourhood. I was also lucky enough to be in an area within New York City that still had a volunteer fire department, the West Hamilton Beach Volunteer Fire Department and Ambulance Corps.

West Hamilton Beach Volunteer Fire Department, taken in 2017.

Whenever there was a fire or ambulance call-out in the area, the siren and horns on top of their firehouse would blare, sometimes muffled by the roar of a passing plane. The horns always caught my attention. Whether I was home or at school, I would stop paying attention to what I was doing and count the seconds until I could hear the sirens of the FDNY leaving their station. If I didn't hear the fire trucks, then I knew that the horns were signalling an ambulance call-out.

The local FDNY station sat on Crossbay Boulevard, not far from my house. Behind its big red door sat Engine 331 and Ladder 173, two big, red, shiny fire trucks with three or four firemen assigned to each. At the back of the firehouse, a small inlet led to Jamaica Bay and docked in the inlet was a fire boat called the *James Hackett*. If the boat was required for a response, the guys from one of the trucks would move onto the boat.

I learned all of this early on. Most of my friends had no idea because they didn't care, but I cared a lot because I had no doubt that one day I would be part of it all.

When I was eight, Mum took me down to the firehouse so I could finally meet some of my heroes. I was given a tour of the firehouse

Dave, aged eight, visiting the fire boat.

22 Emergency Response

and brought down to the fire boat, where I got to try on a full set of fire turnout gear for the first time. The fireman's jacket was bigger than me and weighed a ton, but as far as I was concerned, it fit like a glove. This day convinced me more than any other that I wanted to be a fireman when I grew up.

When I was old enough to cross all the streets in the neighbourhood on my own, I would ride my bike up to Crossbay and sit across from the firehouse waiting for the doors to open so I could watch the trucks responding to an emergency. The firehouse held my heroes, but at first, I was too young and intimidated to visit on my own. As I got older and braver, I'd approach the firehouse and talk to the guys if the door was open.

One of my friends, Mark Mokotoff, who lived across the street, was equally interested in the fire department. He and I would often bike to the firehouse together. One fireman who befriended us was a guy named Freddy Veteree. He gave us lots of time during our visits, teaching us how to slide down the fire pole, roll hoses, clean gear and use the watch-house radio. We learned how his shifts were rostered so we could visit on the days that he was working. I can still remember my excitement the day he invited us to have lunch with the guys in the firehouse — it is a day I have never forgotten.

Two police boats behind the FDNY station — no fire boat anymore, in 2017.

Hating School **23**

My passion to be a fireman grew stronger after a show called *Emergency!* came on TV when I was 10 years old. It was all about the Los Angeles County Fire Department Squad 51 and a newly created type of rescuer called a firefighter/paramedic. Not only did they get to fight fires but they performed rescues and saved people using their medical skills. Johnny Gage and Roy Desoto were the Squad 51 paramedics, and they quickly became my new heroes.

I don't remember what night of the week *Emergency!* was on, but I remember making sure that I was sitting down in front of the TV every week to watch it from the opening credits to the very end. Even today, forty years later, I can recall some of the rescues they performed and fires they fought. *Emergency!* had a massive impact on my life. Now all I wanted to be when I grew up was a firefighter/paramedic!

AS SOON AS I turned thirteen, I joined the Lindenwood Volunteer Ambulance Corps in the next neighbourhood over. As a Youth Squad member, I was taught first aid and CPR, the skills that allowed me to save that stranger's life less than a year later.

A few months after joining Lindenwood, I met Jonah Cohen from the West Hamilton Beach Volunteer Fire Department who told me they also had a youth programme for 'junior' firemen. I visited his firehouse the following weekend and instantly knew that this was where I wanted to spend my time. They had fire engines and an ambulance — the perfect combination for becoming a firefighter and paramedic. At the firehouse, we were called 'buffs' because most of what we did was clean and buff the ambulance and fire engines. I didn't care, I was closer to the action, even if I couldn't be directly involved.

Most of the guys at the firehouse became my mentors and friends, but Jonah is one of the few people I still catch up with when I visit New York. The other volunteers came from all over the area. Some of them were cops or firemen who spent their days off volunteering in their local community. One of them, Kevin Delano, was a professional firefighter and the Chief for many of the years I was a member. He was also one of the 9/11 first responders who later developed acute leukaemia and died, as a result of exposure to materials at the World Trade Center site. Kevin is a true hero in every sense of the word.

I NEVER LIKED school, even as a young kid. I would fake illness on a regular basis from third or fourth grade onwards, trying to avoid at least one day of the most boring place on earth. I would sit in class listening to the planes or the fire engines or the birds; anything other than the teacher at the front of the room. The more time I spent in school, the more I hated it.

Once I joined the ambulance corps and fire department, I realised it was not learning I hated but rather the things we were being taught and how we were taught. I would struggle to tell you how photosynthesis worked five minutes after my science class ended but I would have no problem naming the 206 bones in the human body or explaining how the heart worked. Learning was easy for the things that were important to me.

I was so involved with the firehouse from the age of thirteen that I spent little time with kids my own age during my teenage years. While my friends were out discovering girls, or experimenting with alcohol and drugs, I was happy at the firehouse learning the skills I needed for the life I wanted. I was given the hard word early on: if I got involved with drugs of any kind, I would be kicked out of the fire department, so I was never tempted to try anything, not even pot.

Weeks after saving my first life, I started ninth grade, high school in the USA. Instead of going to the local school, I elected to go to Beach Channel High School, which was located about 10 kilometres from home. Beach Channel was a relatively new school, and it had an oceanography programme which I thought would interest me.

The school was a huge complex on the waterfront of Jamaica Bay. Ironically, it looked and felt like a prison because it was surrounded by high chain-link fences and had security guards at the entrance.

I soon discovered that I hated school as much as ever and only liked my oceanography and scuba-diving courses. The rest of the time I felt like I was in prison. With nearly 5000 students at the school acting as a diversion, I soon mastered the art of skipping classes by sneaking in and out of school anytime I wanted. I'd show up in homeroom in the morning, get marked as present, and then take off whenever I felt like it. By appearing at different classes often enough I somehow managed to get away with it for months.

It would be fair to say that I missed as much of ninth grade as I attended. I became a proficient liar, lying to my parents, my teachers and my friends. When my younger sister started school, Mum went back to work, so most days I would ride the subway system until I got bored and then I would sneak home and watch TV. At dinner, I cheerfully reported that school was great and I was doing well in all my classes. Then several nights a week and on the weekends, I would go to the firehouse.

Of course, being a teenager, I never thought through the consequences of my actions. My mid-year report card was average, but given I had not attended so many classes I was amazed that I was still passing. However, my year-end report card revealed all and judgement day cometh.

The shit hit the fan. Mum and Dad threatened me with everything but death. I was grounded from doing everything I liked, including the fire department. That is when reality struck me. Missing school was one thing but being kept away from the fire department was another. My parents were going to pull me out of Beach Channel and send me to the local high school, but somehow, I don't know how, I talked them out of it. As the summer wore on, I was slowly allowed to return to things that I loved doing, including the fire department, and a part-time job working at a bait and tackle shop on Crossbay. I'd get up at 6am every morning, open the shop, work there for a good part of the day, and then go to the firehouse at night.

By the time school restarted in autumn, I was 15 years old and under no illusions of the dire consequences if I kept skipping school. I would be pulled from the fire department, end of story. This was enough to get me going every day, as much as I hated it. Needless to say, I got to know the school principal and counsellor very well because they were checking up on me daily.

MY JOB IN the bait and tackle shop turned into a weekend job cleaning, filleting and selling fresh fish in a seafood shop owned by Les, from the ambulance corps. One day he and I were heading out for lunch when we came across a horrific car accident. A car travelling on the Belt Parkway had become airborne, crossed the centre divider and smashed into the windscreen of a car travelling in the opposite direction.

There was only a driver in the car that was struck, and even a newbie like me could see that he was dead. His head was partially decapitated, the front half ripped away from his neck. There was blood, bone and brain matter splattered everywhere. It was a gruesome scene and at first it didn't particularly bother me because I was used to seeing and touching fish blood and guts all the time. However, it quickly dawned on me that this was a human being, not a fish, and for a moment I thought I might vomit. I turned my head, took a couple of gulps of fresh air, and then settled down. Les made sure I was okay with what I was seeing, and when he saw I was, we turned our attention to the others.

The car which had flown across the parkway was now sitting on top of the car it had hit, above our head height. While Les made sure that the car was switched off and that there was not any leaking fuel, I started checking the two people in the car. They were both conscious, and although there was some bleeding, nothing looked to be life-threatening. The driver of the car was confused and asking what happened, but I couldn't tell him much.

A few minutes later the fire department and a couple of ambulances arrived. I watched how carefully the firemen respected the body of the dead man as they rescued the two people above. Neither of them needed to be cut from their car, so each was placed on a backboard and brought to a waiting ambulance stretcher. Once the injured pair were removed from their car and sent to the hospital, the firemen turned their attention to removing the dead man from his car. We didn't stick around to watch him be extricated from the vehicle.

I didn't feel like lunch any more, but Les said words to me that would set me up for life. He told me that this was a moment of truth for me, that if we went for lunch and I managed to eat and hold it down, I would never have a problem with anything I saw on the street again. However, if I didn't eat, if I let the accident get the better of me, I'd probably never make it as a firefighter or paramedic. With the stakes being that high, I went for lunch.

Trust me, as I looked down at the cheese and tomato sauce on my slice of pizza, all I could think of was blood and splattered brain, and I felt ill. But as far as I was concerned, my dreams were at stake. Not only did I manage to hold down my lunch that day but I also managed a good sleep. From then on, it was rare that anything I saw on the job affected my ability to eat or sleep.

I REMAINED UNDER the watchful eyes of my parents and teachers as far as school was concerned. By now I tolerated school because I knew that if I skipped, I couldn't do what I loved outside of school. I was now old enough to ride along in the ambulance regularly as well as getting some fire training.

It was around then I learned that success means different things to different people. My parents had a white, New York, middle-class, Jewish vision of success in their minds. For them success for their son meant being a white-collared professional — a doctor, lawyer or stockbroker — making good money, with a nice house, a Jewish wife and 2.5 kids. Someone who was a good provider — someone like them.

When it came to success, I would see 9; they would see 6 — no matter what we said or did I don't think we would ever agree on what success looked like. We couldn't understand each other's point of view.

My parents had absolutely nothing against me wanting to help people and save lives. But their vision was a surgeon working in a hospital, not a fireman racing into a burning building. When we watched *Emergency!*, I wanted to be Johnny Gage, the footloose and fancy-free firefighter saving the injured guy stuck on top of a tall crane. They wanted me to be Dr Bracket, the guy waiting in the ER for the patient to arrive. This difference in opinion caused many arguments. They wanted me to study for exams and apply to college, but I didn't see the point as I didn't need any of that to be a fireman.

My education outside of school never stopped, though. By now, I was riding the ambulance regularly, technically as an observer, but often I was doing patient care, meanwhile studying EMT and paramedic manuals to learn as much as I could.

THREE
EARLY AMBULANCE MISSIONS

A NAKED WOMAN was pretty much a foreign concept to me when I was a teenager. I was so busy with the firehouse that I had never been on a date — I just didn't hang out with girls my age. I went to a few parties but I never felt like I fitted in. There was no internet, so any chance of seeing pornography was limited because buying a 'dirty' magazine from an adult, in public, was hardly worth the embarrassment. Everything I had managed to see was from the occasional *Playboy* magazine left lying around the firehouse.

I was 16 years old when I first saw a naked woman, up close and personal. I was at the firehouse one evening when we got called to a drug overdose at a motel on Crossbay Boulevard. When we arrived, I was the first one off the ambulance and into the motel. The bored clerk at the counter lazily pointed at the room where the patient was waiting. He clearly had done this many times.

I got to the room and looked in and saw a naked woman lying on the bed furthest from the door. I stepped over a couple of things on the floor to reach her; however, once there, my bamboozled teenage brain wasn't sure how to proceed.

Fortunately for her, my training kicked in. Check her breathing, my brain commanded. I tilted her head back and watched her chest rise, totally captivated by the naked breasts in front of me. The good news was that she was breathing, but I am sure I checked her chest movement for longer than necessary. Just then, my partner walked in.

'Is he breathing?' he asked. 'What do you mean, he?' I replied. I mean, she was clearly a woman.

'Not her, you ass, this guy on the floor.' I looked down and was astounded to find that one of the 'things' I had stepped over in my rush to check the naked woman was a naked man.

I was so focused on her that I hadn't even seen him. They were both breathing, but we found out later that they had overdosed on drugs. Soon after, a second ambulance arrived, and we each transported one of the patients to the hospital. My partner told the other ambulance crew to take the woman and we'd take the man. He had decided that I had had enough excitement for one night.

I got a hard time from the guys at the firehouse about that call, and learned a valuable lesson. They hammered into me about how you must thoroughly check every scene for safety before you walk into it. I could have easily walked straight into the room and been attacked by a drugged-up person who thought I was trying to hurt his girlfriend. Lesson learned and experience gained. But hey, I was only sixteen!

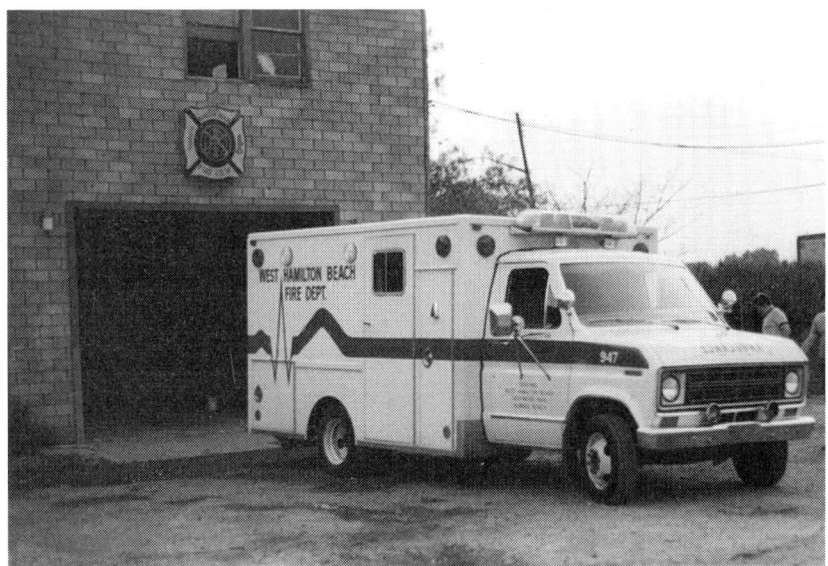

The first ambulance Dave rode on, in front of WHBFD firehouse, 1977.

AFTER I TURNED seventeen, life became easier because I got my driver licence and could drive myself to the firehouse. I could drive my parents' car but not an ambulance until I was eighteen.

My love for the fire department was growing every day, and the uselessness of attending school was becoming more apparent. After finishing tenth and eleventh grades without any problems, my parents and teachers thought I was back on track so they backed off checking up on me every day. As soon as I figured out that I could get away with it again, I took advantage of their trust and skipped a lot of twelfth grade.

The only classes I attended regularly were my computer classes with Mr Harrison, because I loved them, and my maths class with Ms Hempe, because I had a total boyhood crush on her. Otherwise, I skipped school and headed down to the firehouse to wait for the next emergency.

One night, my pager went off at about 1am for an ambulance call-out, so I rushed to the firehouse with my own blue light flashing in the windscreen (which I thought was so cool). We headed to a house where a woman had reported that her sister was bleeding profusely. She was apparently a hysterical wreck on the phone so getting useful

Fifteen-year-old Dave at the WHBFD firehouse.

32 Emergency Response

information out of her was impossible. The dispatchers sent us backup from a city ambulance with a paramedic on board.

On the way to the call, my partner and I discussed everything that could cause a lot of bleeding. The best we could come up with was a stabbing or a person accidentally cutting themselves with a knife.

When we arrived, a woman came running out of the house, still hysterical, and was screaming for us to hurry up because 'the baby was coming'. Well, that explained the bleeding.

By now, I was an advanced first-aider. Delivering a baby was one of the topics taught in training, but everything I knew was from what I read in the books, one classroom lecture, and what I saw in the instructional movie. The movie demonstrated what to do if you were on the scene when a woman was having a baby. I clearly remembered that the best advice was to try and have an experienced doctor with you.

My partner called the dispatchers and told them what was happening and asked for a rush on the paramedics. Dispatch advised us that they were about 15 minutes away. We were on our own for now.

Things were going well. Her waters had broken, but the baby was not crowning much. She was 36 weeks pregnant, pretty much full term these days, but still considered early back then. Her sister had quietened down and was sitting by the mum-to-be's head, keeping her relaxed and breathing evenly. Everything was going to be fine, I thought. We explained that we were waiting for the more highly trained paramedics to arrive before we transported her to hospital, and she was okay with that.

One of the things I had learned in my training was to always wear sterile gloves when delivering a baby. This was back before we routinely wore them on all calls. I cannot tell you how many sets of gloves I went through in her living room. I'd forget I had them on, touch something, get flustered, rip off the now 'dirty' gloves, and put on another pair. At this rate, we were going to run out of gloves before the baby was born.

While we were waiting for the paramedics to arrive, she told us that she felt like she had to go to the toilet and move her bowels.

This was great. I confidently explained to her that this was a normal feeling, as it was just the baby moving towards the birth canal, and she didn't 'really' need to go to the toilet. About a minute later, much to my horror and her embarrassment, the woman delivered a 'number two' right in the middle of her living room on the clean sheets.

'The manual never talked about this!' I thought to myself.

We removed the soiled sheets, going through yet another set of gloves in the process, put down clean sheets, and assured her everything would be okay. We were all relieved to hear a siren getting closer — the paramedics were almost here.

Once the paramedics arrived they assessed the woman and saw that we had things under control. They assured us she was not going to deliver her baby for hours and said it was fine to transport her to LIJ Hospital, her hospital of choice, which was about a 20-minute drive away.

Even though they wanted to clear, I asked if they'd stick with us for the trip. They agreed to put the paramedic into our ambulance with their ambulance following behind. If they got called away to a more urgent job, we'd pull over and the paramedic would jump back into his ambulance and leave us to it. It seemed like a great plan. We bundled her up, got her into the ambulance and headed for hospital. About 10 minutes later the baby who was not going to deliver for hours showed us who was in charge. It was time.

We pulled the ambulance over to the side of the road. The paramedic allowed me the honours and a few minutes later I delivered a baby girl, in the back of our ambulance.

When a birth goes well, as it did this time, the only thing I had to do was gently support the baby's head, as she made her way into the world. Once she was delivered, and a quick assessment showed she was okay, we cut the umbilical cord and wrapped her in blankets. A few minutes later, just before we arrived at the Emergency Room, the placenta delivered. We cleaned the woman up as well as we could and continued to hospital.

We finally arrived and with a huge grin on my face, I rolled the woman and her newborn baby into the ER.

When we got back to the ambulance we found it to be a bloody mess. In all seriousness, the blood from the placenta delivery made a bigger mess than most of the car accidents I had ever attended. But we happily cleaned up and returned to the firehouse.

This was truly a big night for me. Instead of saving a life I'd helped bring a new baby into this world!

IT WAS AROUND this time that my mum uttered hurtful words which have remained with me ever since.

I arrived home one day and was a bit happier than normal. Mum asked why I was so happy and I said: 'I've just been to a great car accident.'

She looked at me like I was a complete nut case and, with a look on her face that I had never seen before, said: 'It's too bad other people have to suffer for you to be happy.'

They were meant as cruel words and they cut me like a knife. It was hard for me to believe my mum said that. Here I was helping people, and I thought that was a good thing. The thing that made me happy was helping them, not the fact that they were in an accident. I never wished for other people to suffer, but if they did, I wanted to be there to help them.

Nevertheless, I have never forgotten her words. They are always a reminder that if I am happy at work it is because someone else is having one of the worst days of their life. It is good to love what I do but I must also be aware of and empathetic to other's feelings.

I DIDN'T WANT to go to college. This was a source of many arguments with my folks during my teens. As far as they were concerned, I needed a college education if I wanted to make anything of my life. I still wanted to be a firefighter/paramedic and I didn't need a college education to do that.

I was naïve and never stopped to consider that when I turned eighteen, I could just move out and do whatever I wanted with my life. In the end, my parents won the battle, and I managed to get accepted into a few colleges. I chose State University of New York (SUNY) at Fredonia, mainly because it was the furthest college from home where I could still qualify for state funding. Fredonia also had a Computer Science Programme, and this was the only subject that interested me.

However, by the end of twelfth grade, I hadn't done enough to graduate high school so I had to attend summer school to graduate. Needless to say, my parents were not happy with me! It was a long hot summer, and I was so stupid that I skipped many of the summer classes, too. Somehow, I got through it and was off to college to study Computer Science soon after.

I needed and wanted to get as far away from New York City as I could. For me, moving away from home was the best way to become the person I wanted to be and escape the iron-fist tyranny that was forcing me to go to college and not allowing me to do what I wanted.

Two days after my eighteenth birthday, I sat in the back of the family car as Mum and Dad made the eight-hour drive to Fredonia. The only thing memorable about the drive was the last few minutes before we drove into the school grounds. For some reason, Dad decided that a baseball cap was not an appropriate piece of clothing for me to wear, and we had a huge argument about it.

Ironically, when Dad retired years later, he took to wearing baseball caps all the time. There were many times I wanted to remind him of that argument. I never did. He probably wouldn't have remembered it, but I have never forgotten it, probably because it was the last big argument of my childhood. A milestone of sorts.

As my parents drove away from Fredonia the next day, I am sure that there would have been a tear in the corner of all our eyes. Regardless of how hard the teenage years had been, I loved them and they loved me. Although, I am equally sure that they would have been doing a happy dance in their seats as soon as their car turned the corner, leaving me behind!

I honestly don't regret all the school I missed because that hasn't made a big difference to my life. However, I do deeply regret all the pain and stress I caused my parents during my teenage years.

Looking back, I also regret missing out on being a teenager, hanging out with kids my own age and experimenting with all those things the '60s and '70s were renowned for. If I had it all to do over again, with the benefit of hindsight, I would have done things a bit differently.

Unfortunately, I rejected most of my parents' advice about anything and everything. I would usually do the opposite of what they said. They told me to study hard. I didn't. They told me how important it was to save my money. I spent it. I'm sure I can squarely blame the fact that I am not a multi-billionaire today at their feet. If they had told me to spend all my money and not save a dime, I would have done the opposite and saved every penny!

The fact that they didn't kick me out of the house, continued to support me and continued to love me unconditionally speaks volumes about the good people they are. All the things I know about being a good person and loving others, I learned from them.

FOUR
UNIVERSITY

THE STATE UNIVERSITY of New York at Fredonia sits in a picturesque village about 5 kilometres from Lake Erie, one of the Great Lakes. I arrived there in late August and the two things that immediately struck me were the clean air and clear skies. Having lived my whole life in the constant smog and light pollution of New York City, I just thought that was how the world was, so it was a revelation to see the abundance of stars in the dark Fredonia sky.

Unfortunately, I soon discovered this starlit beauty came with a massive downside — the winters were brutal. Freezing cold temperatures and a lot of 'lake-effect' snow as the weatherman called it. Simply put, cold air travelling across the warmer water of the lake would pick up water vapour, freeze it and dump it on Fredonia.

We frequently had snowstorms that left half a metre of snow and sometimes twice that. After many weeks of these storms and the snowploughs continually pushing snow to the edges, metres-high snow banks formed along the roads. As it was so cold, they didn't melt until spring, forming tall walls along the roadside.

But on my arrival, the joy of punishing winter snowstorms still lay ahead and my first few weeks were occupied with meeting new people, settling into college life and learning to live away from home for the first time.

Many of the students I met came from small towns from over New York State so everything I found small, they found big. They considered the eight-storey building on campus a sky scraper, while buildings in my neighbourhood were taller. It was fascinating seeing things through the eyes of people who had such different upbringings from mine.

But in other aspects of life, I was the newbie. My life of abstinence in high school bit me in the ass early on. During my second weekend on campus, I decided to hang out in the dorm with other students who were all drinking. I had absolutely no idea what, or how much of anything, was okay to drink, and not wanting to look like an idiot, I chugged down everything that was passed my way.

We all know how that ends, and I learned a few life lessons that night. Firstly, don't mix spirits, beer and wine. Secondly, if you are going to be sick, try to make it to the bathroom not on the carpet in front of everyone you are trying to impress. Thirdly, if you break these rules, don't expect to be a popular kid on campus.

I discovered I was just another computer geek who was never going to be asked to join the cool kids. Had I spent more time at high school socialising, this might not have been such a shock to me. Fitting in proved tougher than I expected but, fortunately, I managed to find an eccentric group of friends to hang out with. Geeks tend to find each other, and although we were from all over the state, with different majors, different backgrounds and different hopes and ambitions, we all needed someone to buddy up with.

And I had a lifeline; now I was eighteen, I could sign up for an Emergency Medical Technician (EMT) course, my first step towards becoming a paramedic. Once a week, I walked to the firehouse in the village, regardless of the weather or anything else going on around campus. I loved the course, and I was good at it, as I knew I would be after all my years of study and practice. I couldn't wait for the day when I could legally ride the ambulance and be allowed to look after patients on my own.

My hatred of schooling followed me to college. I simply didn't want to be there. The only class I enjoyed was computer studies, the others didn't interest me, so I just didn't bother going to most of them. I decided one of the good things at uni was that it was easier to skip class than in high school. No one cared and no one followed up because I was now an adult and I was responsible for myself.

Of course, the chickens came home to roost when my parents saw my grades at the end of the first semester. They hit the roof. At their wits' end, they gave me an ultimatum — if I didn't start passing my courses, they would stop helping me with tuition fees and living expenses, which was fair enough. I had to either do things their way if I wanted their financial support or do things my way and make it on my own. I tried to buckle down in my studies, but it was not working.

During my second semester, I completed my EMT course and got a near-perfect score on all the written and practical exams. It was a huge boost for me to discover I could achieve great things when I applied myself and cared about the subject. Unfortunately, university students who lived on campus were not allowed to join any of the local fire departments.

However, as an EMT student, I did get to do some shifts on the fire department's ambulance. On my first weekend, we got called to a 'wood cutting' accident just outside of town. I was not expecting it to be difficult given all the things I had seen back in New York City. As soon as we arrived, however, I realised that I hadn't seen 'it all'.

A guy had been cutting logs with a chainsaw when it hit a knot in the wood, kicked back and struck him in the head. I had never seen anything like the damage he had to his face. Although it was grisly, it wasn't as bad as the guy with the partial decapitation I had seen years ago. My instructor thought I handled the call well.

We treated him as best we could and then transported him to the local hospital in Dunkirk, where it was decided that he needed to be transferred to Buffalo for higher levels of care and plastic surgery. Mercy Flight, an emergency helicopter service, was called and 90 minutes later they landed nearby to transfer him. We transported him from the hospital to the helipad, and I got my first look at an EMS helicopter. Once I had seen inside and smelt the exhaust as the helicopter started up, I was hooked. Now I wanted to be a helicopter paramedic.

For the next few months, I got to ride the ambulance occasionally, which kept me interested. With a major interstate highway and a lot of wooded areas nearby, we responded to a lot of motor vehicle and hunting accidents, so traumatic injuries followed by helicopter transfers were commonplace. Throughout winter we also responded to numerous weather-related medical and trauma cases. I never knew so many people could have heart attacks shovelling snow.

A CHANCE MEETING with Foster Schucker, the assistant director of the school's computer centre, during my second year at uni, became the catalyst for me moving to New Zealand years later. I met him in the student bar, bought him a beer, and we soon became good friends, and he became my computer mentor.

Thanks to him, I ended up working part-time for the school computer centre and got involved in some interesting projects. By the middle of the second year, I gave up studying full-time and worked at the computer centre. My parents were not thrilled but were happy that I had at least chosen a white-collar profession. It was finally time for me to make it in the world on my own!

DURING THAT TIME, I returned to New York for my older sister's wedding, and I finally got the chance to demonstrate my skills to the family. Midway through the wedding reception Ben, my grandmother's second husband, collapsed. My dad started shouting: 'Is there a doctor in the house?' as he stared at my cousin Alan, who is a doctor, and me, the EMT.

My mum was always the calm one in an emergency, my dad, not so much.

Alan and I started CPR, and I made sure someone rang 911. As we began, the other wedding guests gathered around to see what was happening, but Dad wanted them to go away and enjoy the wedding.

He stood there trying to shoo them off as I heard him muttering over and over, 'He's ruining the *&%$^#@ wedding.' It was bizarre. Alan and I continued CPR until the paramedics arrived and defibrillated Ben, restarting his heart. We broke a few ribs, but Ben survived and lived for many more years.

My dad's father died when I was seven, so I don't remember much about him. Although I didn't know Ben well, I knew he made my grandmother happy, and that was why I considered him part of the family. Saving his life was very special. I suddenly got to see things from the rescuer and family point of view.

Of course, my parents were relieved that we had saved Ben, and I was thankful that they understood my passion a bit more. Although it did backfire somewhat, as it convinced them more than ever that I could become a great doctor if I would just return to studying. For me, finishing university and then going to medical school was akin to a life sentence in prison and I wasn't going to do that.

I wasn't even sure that my memory about the wedding was all correct so at another family event 16 years later I asked Alan if he remembered it. He started chuckling and said: 'Do you remember your dad shooing away the people?'

Years later, when my sisters and I were together, I was telling them about saving Ben and neither of them remembered it. It is amazing how different people remember the same event so differently, if at all.

I RETURNED TO Fredonia and my computer job but never gave up on my firefighter/paramedic dream. I continued my ambulance studies gaining a Level-III EMT. This meant I could insert emergency airways, start intravenous lines, give some drugs and use a defibrillator. A Level-IV EMT was a paramedic, but I ended up moving to Dallas before I achieved this.

A short time later Foster left the university and started working for Fisher-Price Toys near Buffalo. One day he gave me a call and said they were looking for some Unisys Cobol programmers. He thought I could easily do the job and it would be a good move for me. The weekend before the interview he worked with me to make sure I knew all the buzzwords I needed to impress them. Once I accepted their job offer, he spent time making sure I could do all the things I told them I could. We all need friends like him.

When I started working for Fisher-Price, I moved in with Foster, his partner Diane, and her two young children, and I discovered it was lovely to be in a family environment again. They lived in a small community near work and I joined the local fire department in Elma, NY. My EMT qualifications meant I could get on the ambulance immediately, but I had to wait for a firefighter course before I could be a volunteer fireman.

It was great to be back on the ambulance regularly, but I was also enjoying having a well-paid job and being good at this computer stuff. I began to wonder if my dream of becoming a firefighter/paramedic was fading.

My next job was as a programmer at a company based in downtown Buffalo. The company had business interests around the world and one day the boss asked which of us had a passport because they needed someone to fly to London the next day. I didn't have one so missed the opportunity to go, which really annoyed me. That day I went and applied for a US passport to ensure that the next time an international opportunity came around I was ready.

FIVE
DALLAS

IN NOVEMBER 1985, I got a call from a recruiter asking if I was interested in a programming job in Dallas. I was sick of the cold and snowy winters, and so after visiting Dallas for an interview I accepted the job, packed up my things and headed to Texas. I left Buffalo on a cold, snowy, –10-degree Celsius day and arrived to 20 degrees in Dallas. I quickly realised I liked warm weather better than cold!

The move turned out better than I could have expected. My first job there was interesting and challenging, although I didn't get along with my boss, so I moved to a new company as soon as my first year was up. The best part of the job was that I worked with some great people who became my Dallas family.

I loved Dallas, but some Texans didn't love me and my New York accent as much. I learned early on that a 'Yank' was someone from north of the Red River (the river separating Texas from its northeastern neighbours), but a 'damn Yank' was someone from north of the Red River who stayed. I stayed for more than a visit, so I was a damn Yank. Even being nicknamed 'Buck' by a friend didn't help! I liked Texas more than New York, so I acquired a thick skin and put up with all the friendly, and not so friendly, banter.

As soon as I settled in Dallas, I looked for an EMT course and signed up for one at Methodist Hospital where I met one of the best EMS instructors ever. Karen Yates was a skilled instructor who was as passionate about teaching as she was about her emergency nursing. Methodist was home to Careflite Dallas, the EMS helicopter programme serving that part of Texas, which it was great to get exposure to.

As part of my EMT training, I did some ride-alongs with a Dallas Fire Department ambulance unit. I did a few shifts and found a renewed passion for becoming a firefighter/paramedic. Since there were not any volunteer ambulance or fire departments in Dallas City, I had to find new ways to help.

I joined the Red Cross Disaster team which helped people with everything from their home burning down to being destroyed by a tornado. We also set up a refreshment area for firefighters at the scenes of large fires. I wasn't called out often, and it was really satisfying work, but there was not much adventure and I wasn't saving any lives — it was good, but not good enough!

I also became a volunteer at Parkland Hospital working in their Emergency Room every Friday night from 11pm to 7am Saturday morning.

I was exposed to a lot of trauma injuries at Parkland ER including many stabbings and shootings every week. Since I was an EMT, I was more involved than some of the other volunteers in the ER. During a shift, I would do anything from taking blood pressures in the waiting room, helping bandage in a treatment room, cleaning up a bloody trauma room, to doing CPR. Whatever was needed I would do. I was not quite living the dream but being this involved was great.

Often a call would come in to the ER that a helicopter was about to arrive on the roof. Sometimes, the patient was brought straight down by the helicopter paramedic and nurse or the medical team would head to the roof and start treatment as the patient was taken off the helicopter.

There was always a sense of anticipation when a helicopter was approaching as they inevitably carried the sickest or most critically injured patients. I was in awe of the helicopter teams when they came in, their flight suits and helmets adding a real mystique, and I started wondering what it would take to become a flight paramedic. At the very least, I knew I would have to give up my day job and find a full-time job as a paramedic.

Every Saturday morning when I got home, I threw my bloodstained clothes into the washing machine and had a long shower to wash away the shift. I always had trouble falling asleep that night, not because of what I had seen or done, but because it was a weekly reminder that I was not doing what I wanted to do with my life. I was frustrated, but still not brave enough to make the jump away from a well-paid IT career.

IN JANUARY 1988, one of the major TV networks was launching a new TV show called *48 Hours*, hosted by Dan Rather, the news anchor for the *CBS Evening News,* which was going to focus on 48 hours at a specific location.

For the first episode, they chose Parkland Hospital, as this was where President John F. Kennedy was brought after he was shot in 1963. Dan was the young reporter who covered the story for CBS, and it marked the beginning of his rise to stardom.

The producers of *48 Hours* were quite intrigued by my double life; a transplanted New Yorker who was a computer programmer by day and who volunteered to immerse himself in the blood and excitement of Parkland ER every Friday night. They decided I would be a great interviewee for the show and filmed me working the day job and then working in the ER.

Finally, my big moment came to meet Dan Rather. In my mind, meeting Dan was akin to meeting the President. Despite his small talk and easy conversation to calm me down, I was so nervous that I became a sweating, tongue-tied idiot. I did so badly with the first interview that they had to reshoot it. I didn't do much better the second time and, unfortunately, unbeknown to me, they didn't show bad interviews on air.

I told everyone I knew, and many I didn't know, that I was going to be on the inaugural episode of *48 Hours*. I sat down expectantly only to see a few shots barely recognisable as me in the background. I was crushed.

I am so thankful that Facebook did not exist back then! Although I had some very supportive friends, some were more supportive than others! The funniest and most in-my-face message came from my workmate Al who sent me a message saying: 'Unless you were the guy being interviewed in the clown suit, I didn't see you on TV last night — oh yeah, the clown didn't have a New York accent. Ha ha.'

I wanted to crawl into a hole and never come out. I was so damned embarrassed, and there was nothing I could do about it! That night I vowed that I would never appear in front of a TV camera again in my life, no matter what. Little did I know.

AROUND MY TWENTY-SEVENTH birthday, I paused to think about where I was in my life versus my life dreams and goals. I certainly wasn't a fireman. My 'dual life' was working well. I liked my job, plus the volunteering I was doing was satisfying, even if it fell well short of being a fireman. Life was not too bad — but it could be better.

Even so, I knew I was settling for the life I had instead of creating the one I had wanted so long, the life I had dreamed of since I was eight years old. But how to change it was too hard to deal with, so I did what every unreasonable person would do at this point — nothing.

In September 1989, I read that the Dallas Fire Department was going to recruit firefighter/paramedics for the first time in many years, with the first intake early in 1990. It was a bit contentious because they now required new recruits to become firefighters and paramedics. Many people only wanted to be a firefighter so they didn't like this, but here was my dream job being waved in front of me.

Yet, what should have been an easy decision for me became a huge dilemma. I enjoyed the challenges and travel of my programming job. Being a fireman certainly didn't pay as well, and I had settled into a comfortable lifestyle. Was I willing to trade it in and apply?

The more I thought about it the more self-doubt crept in. What if I applied and didn't get accepted? What if I didn't pass the medical or physical? What if I got into the training school and then flunked out? How would I pay for my new car on a fireman's salary? What would Mum and Dad say . . .?

After a lot of internal debate, I decided that this was my time to put up or shut up and I mailed in my application. Although, after the Dan Rather TV debacle, I didn't tell many people that I was applying. If I wasn't successful I wanted to hide my shame!

A month after applying, I got a letter scheduling me to take a civil servant exam. Passing this was required for any city job. Thankfully, missing out so much high school was not a hindrance and I easily passed. The next step was a medical exam. No problem. Then another written exam, this one specifically about becoming a firefighter/paramedic. I passed this exam with flying colours, too. The final hurdle was passing a physical exam which I did without a problem.

I was incredibly excited. My dream job was suddenly so close I could practically reach out and touch it. All I had to do was wait for the city to announce a start date for the course, but that announcement wasn't expected until the new year which was months away.

Even though it was pretty much a done deal, I only told my closest friends. I didn't think everyone would understand or support my decision. I decided to leave giving notice to work and telling my parents to closer to the date — no sense in starting that argument till I had to!

SIX
FIRST TRIP TO WELLINGTON

MY COMPUTER JOB was not inspiring me any more, but it paid well and I knew I had to hang in until the firefighter course began. Then fate stepped in.

One evening in October 1989, I received a surprise phone call that changed my life forever. John Cunningham, an 'Englishman' (he was a New Zealander) from Unisys New Zealand rang and asked if I'd like to move to Wellington for a one-year programming contract. A project was underway to write a new tax system and they needed someone with my skills to help.

At first I thought it was a prank call set up by my friends but he assured me it was real. They had been searching for someone with specific skills and Foster at Unisys World Headquarters had recommended me. I was chuffed that my skills were considered good enough for the project, and I knew that Foster wouldn't risk his reputation if he thought I couldn't deliver. As it was so out of the blue, John said he'd give me a couple of days to think about it.

I hung up the phone in disbelief. My lifelong dream of becoming a firefighter was about to come true, but in my most adventurous dreams I had never thought of leaving the USA. Now I had the chance, I didn't know what to do.

I didn't know much about New Zealand except that it was a small country with few people and lots of sheep. I had recently read about

France blowing up a Greenpeace ship there and something about a nuclear-free zone. I could find it on a world map, mainly because I was an avid scuba-diver and it was close to the Great Barrier Reef in Australia, which I had always dreamed of visiting. Being 1989, it was pre-internet so I couldn't do more than check my Encyclopaedia Britannica.

I called Foster and we had a good chat about the offer. He thought it was a great opportunity and I should go for it — some overseas work would look good on my CV. He had visited New Zealand for a project a couple of years before and said it was a great place.

I liked adventures, but New Zealand? Forget living in a different time zone to New York, this was in a different hemisphere. It was a world away from everything and everyone I knew. With my dream of being a fireman so close, why would I even consider this offer?

When John rang a couple of days later, I told him I was interested but wasn't sure about moving to New Zealand. He suggested I visit for a couple of weeks to see if I liked it. If it didn't feel right, I was under no obligation to take the job.

What?! They were offering to fly me all-expenses paid in business class halfway around the world to somewhere I'd never have visited on my own. How could I refuse?

Like most Americans, I only got two weeks' annual leave, so it meant using most of it for this one trip. Thanksgiving holiday was coming up, so that would save me two days. My boss had no issue with me taking the time off, but she was bemused by my sudden interest in New Zealand and didn't seem convinced when I denied that it was for a job interview. She was a smart woman and wished me luck for my vacation.

I let John know that I would love to come out for a visit but reminded him that I wasn't sure I'd take the job. He was pleased I was coming and wasn't worried that I wasn't convinced — yet!

The last thing I had to do was disappoint my mum. I was meant to go to New York for Thanksgiving so I casually mentioned to her that I wouldn't be coming because I was going to New Zealand for a job interview 'but don't worry because I doubt I will even take it'.

She was thrilled and proud that I had this opportunity but wasn't so sure that I wouldn't take the job.

I WAS AS excited as a kid in a lolly shop when I arrived at Dallas Fort Worth International Airport for the first flight of three to get to Wellington. I felt like a fraud because I had no real intention of taking the job, but I had been honest with Unisys and if they wanted to pay for me to come who was I to argue?

My longest plane flight to this point had been from Dallas to Washington State, about a six-hour flight, so I was unsure how much I'd enjoy the 12-hour flight from Los Angeles to Auckland. I discovered that flying business class made the long flight quite bearable.

The sun was just rising as we landed in New Zealand. I was sitting in a window seat taking in all the beauty as we crossed over the Coromandel Peninsula on the approach to Auckland. Some of the greenest hills and pastures spread out below me, and the sparkling blue water was lit by the glow of the rising sun. I was watching the dawn of a new life, but I wasn't clever enough to realise it at the time.

The flight from Auckland to Wellington took about an hour and I stared out my window for the entire trip. It was a clear day and I watched the mountains turn into flat areas and then back into mountains as we flew south. As we descended, it started getting a bit bumpy, just as the captain had warned during his last announcement.

The landing at Wellington Airport was impressive and a bit scary. As we approached I was staring down the runway from my window. I thought this was unusual because the pilot should have the runway out his front window. When the plane was a few metres above the runway, the plane centred as the wheels touched the ground. It was obvious that he had done this before. In the USA, passengers would have applauded this incredible landing. Here everyone around me seemed like they had experienced it before.

John was waiting for me in the terminal. When we headed to his car he told me he would drive as I went to get in the wrong side of the car. Everything was familiar but different at the same time.

It was a beautiful sunny day and we took a stunning scenic route around Wellington Harbour to the city; but the wind was certainly persistent.

That afternoon I wandered the streets and the waterfront. I found it strange that the shops were closed because it was a Sunday, but plenty of people were about enjoying the beautiful day.

The next morning, I walked a few hundred metres from my hotel to the Unisys building. Walking down a street in Wellington was vastly different to New York or Dallas. The odd accents, cars driving down the wrong side of the road and no sky scrapers.

The next two weeks were more about convincing me to take the contract than about interviewing me or doing any work. I was introduced to everyone from the tea-lady to the CEO, and they all were friendly and welcoming.

I spent time with different people each day, and would go out for nice lunches and dinners with different people. They didn't spare any expense in trying to convince me to join them.

People kept telling me that the South Island was even more stunning than the North Island, but I found that hard to believe. Someone suggested that I take a ferry trip down to the South Island on Friday afternoon, drive myself to Christchurch and then fly back on Sunday afternoon and join them for a barbecue when I returned. It sounded like a great plan.

On Friday morning, I rang my parents because it was Thanksgiving Day back in New York. The family was together and Mum told me I was missed. She was pleased I was enjoying myself but I suspect she wished I didn't like it quite so much.

That afternoon I boarded the ferry for the three-hour trip to Picton. I left a cool, grey day behind in Wellington but the weather improved as we crossed Cook Strait and entered the Marlborough Sounds. It was one of the most beautiful places I had ever seen in my life. The ferry slowly made its way through crystal-clear water surrounded by mountains and seaside homes and boats whizzing by filled with people fishing, skiing or just enjoying the tranquillity. Since I still thought this was going to be my one and only trip to New Zealand I was glad to be experiencing this.

I was a bit nervous about driving on the 'wrong side' of the road but was sure I'd manage it okay until I found that the car rental agency had given me a manual car. It was bad enough driving on the other side so I didn't want to deal with a stick shift. I went back and asked if they had an automatic available. She just stared at me and said: 'This is Picton.' Finally, I realised that this meant no. It was a manual or nothing.

I headed south towards Christchurch leaving chunks of metal from my bad clutching along the way. Before long I reached the scariest piece of road I had ever driven on: a narrow one-lane bridge with a railway track across the top of it. This was the main highway of the South Island!

Somehow, I made it safely to Kaikoura, a beautiful coastal town about halfway between Picton and Christchurch. I had been told to grab crayfish for dinner from a roadside stand, and it was as tasty as I had been promised.

As I went to sleep that night, listening to the waves lapping the nearby shore, I reflected on the past week. I hadn't come to New Zealand with any intention of taking the job, but I was not so sure any more. It was a lovely place with lovely people, and I had been handed the opportunity to move here. If it wasn't for the offer from the Dallas Fire Department, moving here would be a no-brainer.

I woke up to a gorgeous sunrise and then continued driving south. The driving was easier than I had expected it to be although I'm sure I pissed off many drivers by going slow on some of the winding bits. I was courteous and pulled over to let traffic pass when it was safe and was impressed by the friendliness of some of the other drivers giving me the 'peace sign' as they passed by. I was disappointed to find out later that it was not a peace sign — the two-finger salute in New Zealand is the equivalent of the middle finger salute in the USA — there were a lot of things I didn't know!

After arriving in Christchurch, I wandered around the outskirts of the city centre. I found a park where two teams in white were playing what looked like a strange version of baseball. I watched for a while and got chatting with a local who explained that the game was called cricket, and, next to rugby, it was the most popular spectator sport in New Zealand. I didn't have the courage to ask him what rugby was!

When I walked back to the city, the streets which had buzzed with people a little while ago were now a virtual ghost town. The shops had closed for the day; they were only open until 1pm on a Saturday and would be closed all day Sunday. This was certainly different to the States.

I drove around Christchurch for a while and eventually found myself back near the airport. I decided that if I was going to explore a city it should be Wellington, not Christchurch. I enjoyed talking with the people while I waited for my flight and was impressed by how friendly and chatty everyone was. After nearly a week in the country, I found it easier to understand the Kiwi accent, but the locals still struggled to understand me.

The flight back to Wellington was spectacular as we flew over the stunning Kaikoura Ranges. As we approached the top of the South Island I could see the magnificent land and seascape of the Marlborough Sounds that I had sailed through two days ago. It was a truly beautiful country.

The landing in Wellington was smoother than the one I experienced the week before, and it was a lovely still evening when I returned to my hotel. No one was expecting me back until the next day so I walked around the city, taking in the sights and sounds while I considered my options, until dark.

By the time I went to bed on Saturday night, I realised why they had been so willing to fly me out. I was falling in love with New Zealand, and Wellington felt like a place I could easily live. The work would be exciting, the people were nice and it was only for a year. But, this was not my dream job. That was waiting for me back in Dallas.

Another thing that had me bemused, if not a bit uneasy, was how excited people seemed to be about the upcoming start of TV3, the nation's third TV channel. Could I really move to a place that was excited about a third channel?

The next morning, I awoke to a different Wellington. A massive windstorm was rattling my windows and for the first time I saw horizontal rain. The mountains that I knew surrounded the region had disappeared. It was also much colder so I assumed, given the weather, the barbecue would be called off. A quick phone call put that idea to rest; rain or not this was springtime and the barbecue was on!

I was driven out to Whitby, a suburb about 30 minutes north of Wellington. I met the families of the people that I had been with for the last week, and they were all warm and friendly, encouraging me to move out for the year. The barbecue was well sheltered and Nigel Miller the host wasn't going to let a boring old southerly blast stop him from entertaining.

It was a great afternoon and I enjoyed it immensely. By the time I returned to the hotel that night I was almost convinced I could live here. But was I willing to put my dream of being a fireman on hold?

Week two flew by with bits of work, lots of food and drink, and some conversations about whether I would take the job. On my last day in their office, I was handed a contract which was quite impressive. They were offering me a great salary, a car, four weeks' holiday and two business class trips back to the USA. How could I say no?

When they dropped me at the airport, they seemed convinced that they'd see me again soon. I was close, but not 100 per cent sure.

SEVEN
CROSSROADS

WHEN I GOT home a letter from the Dallas Fire Department was waiting. It was official; my firefighter training was starting in March.

What a quandary! On one hand, I had the offer of my dream job and on the other, a once-in-a-lifetime opportunity to live halfway around the world. My good friends had different opinions on what they thought I should do. I hadn't told my parents that I had been accepted into the fire department and I didn't think this was a good time to tell them or ask for their opinion.

I called the Dallas Fire Department and they told me that I could defer my entry for 12 months before I'd have to re-sit all the tests. This gave me some much-needed flexibility.

After weighing up the pros and cons, I decided that I'd take the contract and delay my firefighter dream for a year. Several factors swayed my decision: Unisys was paying for me to move countries; I could fulfil my long-held dream of diving at the Great Barrier Reef; it was only a year so even if I hated it, I could survive it. I also suffered from the fear of missing out; if I didn't take the contract I'd always wonder what it would have been like and whether I had missed out on the best opportunity of my life.

Once I decided that I'd take the contract things moved quickly. First, we had to finalise the deal. These were the days before email so we negotiated by phone and by sending an 18-page fax back and forth. Once I had a signed agreement I gave my three-week notice at work and started to prepare for my big move.

Over the next six weeks, I packed up and gave away or sold just about everything I owned. I had two dogs who would have to spend nine months in quarantine before they would be allowed into New Zealand, so bringing them with me was not viable. I was extremely lucky and found a loving family to adopt them both meaning they would be able to stay together.

One of my sisters bought my car so I drove for 24 hours from Dallas to New York to deliver it and catch up with family and friends to say my goodbyes. My mum was not delighted about my move to New Zealand because, despite my assurances, she doubted that I'd stay there for only a year. Mothers really do know best!

The best advice I received was from my friend and mentor Foster who, on the morning I left for New Zealand, called me and said: 'Greenberg, whatever you do, don't screw up.'

I spent my last day in Dallas running around saying goodbye to friends. I was not counting on the long rental car queue at the airport and I managed to miss my flight to Los Angeles. This truly was a screw-up. I had one important job that day, to get on the plane, and I messed that up badly!

Luckily, I was a frequent flyer with the airline and they managed to put me on their next flight to Los Angeles, in first class, no less. When we landed at LAX there was a gate agent waiting who drove me through the airport to get me to my next flight on time.

Had I missed the flight to Auckland, I'd have missed my first day of a new life. Luckily, problem solving is something I have always been good at, even when it's me who creates the problem!

EIGHT
THE MOVE TO NEW ZEALAND

I ARRIVED BACK in Wellington on the morning of Sunday, 14 January 1990 and I have never looked back.

I was contracted by Unisys to work on an Inland Revenue Department (IRD) project that was headed by an American company, Anderson Consulting. This meant that I had to work and please bosses and supervisors from three different organisations.

The difference between the culture of the IRD employees and the Anderson employees was stark. The IRD staff were relaxed Kiwi civil servants, who took their morning tea at 10am, lunch at noon and afternoon tea at 3pm, regardless of what was going on. The Anderson consultants started work early, ate lunch at their desk and finished late.

The Anderson men wore crisp suits, highly starched white shirts and power ties, while the IRD men wore a collared shirt, shorts, long socks and sandals — the standard civil service outfit of the time.

My work culture leaned more towards the Anderson way, although I never put starch into a single shirt I owned. Like me, most of the Anderson people were young, single and from overseas. We worked long hours, drank too much coffee, and partied on the weekends. Many Friday evenings I'd walk over and have drinks with the Unisys team and then meet up at a bar with the Anderson crowd later.

Outside of work I was experiencing many culture shocks, too.

I had to learn the metric system, get comfortable driving on the other side of the road and find my way around a new city and country. I also had to tone down my New York and American habits and mannerisms. In general, Kiwis are much more laid back than Americans and I quickly learned that calling a spade a spade was not always the best thing to do because it would offend people.

In 1990, supermarkets in Wellington city closed at 5.30pm each day, with one late night a week when they stayed open until 7pm. They opened from 8am to 1pm on Saturday and were closed on Sundays. A far cry from the 24/7 shopping available in Dallas. Given my long work hours, and the hangovers on Saturday mornings, I had to do most of my grocery shopping in the local dairy and occasionally drive out to a supermarket in Lower Hutt or Paraparaumu on the weekend.

My American accent was a blessing and a curse. Kiwis seem to love Americans, and the American accent, so simply saying 'hello' was enough to get a conversation going. However, it was very different from the Kiwi accent, and at times I felt like I was speaking a foreign language.

It was similar to when I first moved to Texas. I had to learn a whole new set of lingo and slang, and which words had different meanings in the USA. It took me months to realise that 'al-a-min-e-um' was not a strange material from another planet, but simply 'aluminum'.

On one of my first weekends, I was at a party when one of the Australian guys told me he had to head outside and chunder. That sounded like fun so I went with him and quickly learned that chunder was, in fact, vomit. This was no fun at all and I ended up chundering alongside him. On the positive side, I now understood what the band 'Men at Work' were singing about in their song about the men down under.

Ironically, I ended up renting a house in a suburb called Brooklyn. It was partially furnished, but I needed a bed and bought one from a shop in Thorndon Quay. The owner was a nice guy who, when he heard I was new to Wellington, invited me to a party on the following weekend. Some of the people I met at that party helped shape my New Zealand experience.

Life was cruising along nicely, I was truly loving Wellington and the New Zealand lifestyle. I was living in a city surrounded by the sea so it seemed natural to start scuba-diving again. I joined a local dive club and continued to gain more advanced certifications. The first holiday I booked was a diving trip to Cairns, Australia, and my dream of diving the Great Barrier Reef became a reality.

My work was going well, but ructions were happening with people way above my pay grade. The IRD contract was huge for Unisys and they were willing to do whatever it took to ensure that the project was a success. Much to my delight and surprise, they convinced Foster to temporarily leave his family behind and come to New Zealand to keep the project on track. It was great flatting and working with my friend and mentor again.

In June 1990, I faced another major dilemma. Unisys offered me a one-year extension to my contract and I wanted to take it. Extending the contract would mean losing my spot in the Dallas Fire Department employment queue. After a lot of internal debate about what I wanted to do with my life, I accepted the offer.

Once I knew I'd be in Wellington for another 18 months I decided that it was time to get back into the emergency services. I considered becoming a volunteer fireman, but I'd have had to move out of Wellington City and live in an area with a volunteer brigade. The nearest brigade was only 10 or 12 kilometres away, but I was enjoying being in the city, so I decided against it. It seemed now I was willing to let a mere 10 kilometres stand between me and my dreams.

I was still a registered EMT back in Dallas so I approached the Wellington Free Ambulance (WFA) service to talk to them about joining as a volunteer. I met with a guy who obviously wasn't a fan of Americans. I was told that being an EMT in Texas 'means nothing here' so if I wanted to join, I'd have to start with a basic first-aid course and work my way up. If I was lucky I'd be riding the ambulance in a year or two. This wasn't going to work for me either.

My last stop in May 1990 was Life Flight, the operator of the Westpac Rescue Helicopter. I knew getting a position on the rescue helicopter was a long-shot but I have always believed that if you don't ask, you don't get. I met the General Manager, Peter Mairs, who was cordial and polite but told me they had a long line of people wanting to volunteer on the helicopter, and because I only had time available on the weekend it meant I could not help during the week, when help was most needed.

While I was there, Peter Mairs told me about how the Trust was founded. Standing on the shore of Seatoun Peter Button had helped rescue survivors after the inter-island ferry, the *Wahine*, ran aground in Wellington Harbour, during a cyclone in 1968. He could see that a helicopter would have been able to rescue people in the surf, and save some of the 51 people who perished. On that day he thought, 'there has to be a better way,' and the idea for a rescue helicopter service was born.

Wellington's first-ever rescue helicopter service was started in 1975 by Peter Button and Russell Worth, a local neurosurgeon, who both passionately believed that the service was required. Clive Button, Peter's son, worked with his dad on the helicopter in the early days and the father and son team saved many lives together.

Tragically, on 20 November 1987, Peter Button and two other men, Ronald Woolf and Dion Savage, were killed when their helicopter crashed just north of Wellington. The helicopter had been on a non-emergency flight when the police requested that Peter help locate an escaped prisoner; during this search, the helicopter hit high voltage power lines.

Although Peter Button perished that day, his dream of a Wellington-based rescue helicopter did not die. The rescue service continued to grow and his legacy lives on today.

WHILE I WAS at the hangar, I got a whiff of the Jet-A1 fuel and decided that working on the rescue helicopter was what I wanted to do next! He said it wouldn't be a problem if I stayed in touch with him but I shouldn't get my hopes up about getting on the helicopter.

I ended up becoming a volunteer first-aid and CPR instructor for the New Zealand Red Cross. This gave me the opportunity to keep my skills fresh and meet some new people. One of the people I met along the way arranged for me to do a ride-along with his flatmate who was an ambulance officer at WFA. I arrived at the WFA station at around 6pm and met Dean Voelkerling, who showed me around the station and our ambulance for the night. I asked where his partner was and he told me I was his partner for the night. I was surprised to learn that ambulances in New Zealand were routinely sent out with only one ambulance officer who had to drive and look after a patient at the same time.

Our first job of the night was responding to a person struck by a car on Willis Street. There was a lot of blood, broken bones and upset people. As I sat in the back of the ambulance with the patient on the way to hospital, happier than I had been in ages, my mum's words from years before rang in my ears — 'It's too bad other people have to suffer for you to be happy.' My childhood dream of being a firefighter had never diminished and it was true, I was happiest when I was helping others. I didn't wish anyone harm, but if they needed help, I wanted to be there to offer it to them.

I managed a few more ride-alongs with Dean over the next few months and was ecstatic being back around the life I loved so much. I knew in my heart that I really wanted to get onto the helicopter so I touched base with Peter Mairs at Life Flight every couple of months. He was always friendly and polite but there were still no opportunities and he didn't think this would change anytime soon.

NINE
MONEY AND YOU

I MET MY friend Carolin early in 1990, and 27 years later she remains one of my closest friends. When we first met, she had an administrative job for one of the big banks and hated every minute of it. Caro had big dreams of what she wanted her life to look like, none of which included working at a bank. Her story had a familiar ring to it!

In late 1990, she attended a self-development course, called 'Money and You', which inspired her to make some dramatic changes in her life. Straight after the course, she started to plot how she would resign from her corporate job to concentrate on becoming a massage therapist.

The change was so dramatic that I wondered if she had been brain-washed and was now part of a cult. However, she spent a lot of time convincing me that this was not the case; she simply had been shown ways to empower herself and do what she really wanted to do with her life. More importantly than just telling me, she was demonstrating the changes in her own life.

I liked the changes I saw in her and early in 1991 I decided to attend the course myself. Much to my disappointment the course was more about 'you' than 'money', but it was a life-changing weekend for me.

MONEY AND YOU was one of several self-development courses run by a US-based company, in the 1990s. I did several of their courses and each gave me different life skills, most of which I still use today.

One of the directors and instructors of the company was a guy called Robert Kiyosaki, a larger than life figure who ended up having a dramatic influence on my life. He had been a pilot in the Vietnam War and used his helicopter stories as metaphors for business and life, something I have successfully added to my own repertoire.

Today, Robert is a *NY Times* best-selling author best known for his 'Rich Dad, Poor Dad' series of books. The first book he wrote, which still sits proudly on my bookshelf, was called *If You Want to be Rich and Happy Don't Go to School*. As someone who hated school so much I wish that book had been around in my teens!

The main thing the course did for me was open my eyes to the fact that chasing money, instead of my dreams, wouldn't result in the life I wanted. I walked away from the weekend knowing that I absolutely had to find a way to achieve my emergency service dream.

Just as importantly, the courses brought some incredible people into my life. Today, more than a quarter of a century later, many of my closest friends are people that I met through Money and You.

The following week, I went to my Unisys bosses and told them that I needed to figure out a way to alter my work schedule so that I could have at least one weekday available for volunteer work. They were happy for me to do this, if I put in enough hours and kept IRD happy.

With that handled, I called Peter Mairs at Life Flight to tell him that I could now volunteer during the week. He was pleased with that and told me that there had been a few changes and there might be an opportunity for me to become a volunteer.

My stars were aligning! We arranged for me to meet John Goldswain, the Trust's Operations Manager, one evening the following week.

I managed to stay cool during the phone call but as soon as I hung up I jumped around with excitement. I was convinced that this was my opportunity to get back into the emergency services, in a more exciting way than I could have ever dreamed.

I met with John the following Tuesday night. John and I shared our 'stories' during what was to become the first of thousands of chats over cups of coffee. He was from the UK where he had a varied and interesting background as an electrical engineer, commercial diver and Thames river boat captain. He met his wife Wendy, who was a Kiwi, during her OE to the UK and followed her back to Wellington when she returned home. He had been a volunteer on the rescue helicopter for a couple of years.

He was happy with my prior emergency service experience and said that it would be worth me coming in for some training. If things went well and the chief pilot was happy, they would take me on as a volunteer. He told me he would be in touch and arrange some weekend training for me. I walked out convinced that I had found my new emergency service home.

John didn't ring me that week which I found extremely frustrating. On the weekend, I parked up near the airport runway and watched enviously as the helicopter took off and headed around the south coast. I didn't know where they were heading, or what they were doing, but I knew I wanted to be on that helicopter.

When he didn't ring the following week, I started to panic. Had they brought on another volunteer? Did he think I was unsuitable? My mind started filling in the blanks with answers I didn't like.

I didn't want to seem like an impatient pushy American, but I also didn't want to miss this opportunity. Finally, I mustered up the courage to call John and ask if he had any idea when my first training day might be. He said that he had been busy and had forgotten, but invited me to come in at 10am the next Saturday.

I arrived at the hangar promptly at 10am and John introduced me to Toby Clark, the chief helicopter pilot, and Dave Ross, one of the helicopter volunteers.

Toby was a tall, handsome, blond-haired and good-natured guy who oozed confidence. If someone asked me to describe what a rescue helicopter pilot might look like, I would describe him. He was an extremely experienced helicopter pilot who had flown many types of helicopters in all sorts of operations, including venison recovery in the '80s.

Dave was a tall, solidly built man with a firm handshake. He looked like he could be intimidating if he wanted to be, but was all smiles and welcoming towards me. He was a policeman who had been a community constable near the airport when the rescue service began. Since he lived and worked nearby, he stopped in and offered to become a volunteer crewman, which was gratefully accepted. His current police role was as a driving instructor, a Monday to Friday role, which meant he was available to volunteer on the helicopter every weekend.

The four of us sat down and had a chat about our lives, our backgrounds and the rescue helicopter. Although my previous emergency service experience was a major bonus, having no prior aviation experience made it clear to us all that I had a lot to learn before I could be a useful helicopter crewman.

I was told that Dave, who had considerable experience training people, would do most of my training each weekend. John would train me to use the rescue winch as well as in some of the other helicopter flight roles and eventually, when I was ready, he and Toby would sign me off on helicopter skills. The rest, as they say, is history.

Dave was not only an excellent trainer, but he became a close friend and was my best man at my wedding a few years later. He thought that having two Daves on the helicopter was not on, so he declared that he would be BigD and I would be LittleD going forward. Although we still call each other by these nicknames, they never caught on with others.

TEN
WELCOME TO LIFE FLIGHT

MY TRAINING BEGAN in earnest the next weekend. After we had a look through the helicopter, Dave explained how we fitted in with the emergency services and the other rescue helicopters around the country.

Unlike the USA, the New Zealand air ambulance and rescue system is a network of rescue helicopters and fixed-wing aircraft run by private and charitable organisations. As an American, I was astounded to hear that anyone using the service in New Zealand was not charged for their flight. A combination of government money and charitable funds, including money from sponsors, is used to pay for the flights. It does not cost the 'end-user', the patient or person being rescued, anything.

Life Flight is the charitable trust that runs the Wellington-based Westpac Rescue Helicopter. In 1991 the helicopter used was an AS350B helicopter, nicknamed a 'Squirrel' in New Zealand and an 'AStar' in the USA. The helicopter was a single engine helicopter which could carry one stretcher patient.

The Trust was established in 1982 to keep funds donated to the service separate from Peter Button's commercial aviation business. Westpac Bank is the main sponsor of Wellington's rescue helicopter service which gives them the naming rights as well as the distinctive red and yellow paint job on the helicopter.

The Trust also runs an aeroplane which is used to transfer patients between hospitals around the country. The planes are known as a 'fixed-wing' air ambulance, thanks to their rigid wings, as opposed to the rotors of a helicopter. Dave told me that the aircraft were simply referred to as 'the helicopter' or 'the fixed-wing'.

I was surprised by the variety of missions that the rescue helicopter responded to, everything from car accidents to transferring patients between hospitals and searching for people lost at sea.

The helicopter's main operating area was the Greater Wellington region, which includes the Kapiti Coast, Wairarapa and the Marlborough region at the top of the South Island. The fixed-wing is used to transfer patients from regional hospitals to Wellington, or from Wellington to Auckland or Christchurch for higher levels of care than they could get locally.

The Trust didn't own any of the aircraft; it leased them from a company which also supplied the pilots and all the maintenance. The ambulance officers who flew on the helicopter worked for Wellington Free Ambulance, and the doctors and nurses who flew on any of the aircraft worked for Wellington Hospital. The only person from Life Flight, even if they were a volunteer, was the crewperson.

A map of New Zealand laid over the USA to show the size difference.

Dave begins his career with Life Flight.

MY FIRST FEW weekends were full of ground training and cleaning the aircraft, but three weeks after I started, I had my first flight.

One evening, the helicopter trailer was carefully towed out of the hangar and onto its assigned spot on the tarmac. The sun had set about an hour earlier so it was dark enough for a training flight using the Nitesun, the helicopter's powerful 30-million candle-power searchlight. It was a warm evening with a slight northerly breeze, far different to the gale that had been blowing earlier in the day.

I have always loved flying and was extremely excited about my first helicopter flight over Wellington and only my second helicopter flight ever. My excitement was tempered by the nervousness I felt knowing this was my chance to show Toby and John that I had taken in all my training over the last few weeks and that I had the potential to be a good helicopter crewman. If I screwed up tonight it might be both my first and last flight with them!

I settled into my seat in the back as John did a safety walk around and Toby started up. I felt my pulse begin racing, and my smile growing, when I got a whiff of the helicopter exhaust as the burning Jet-A1 fuel began flowing through the engine.

Welcome to Life Flight

The trailer began gently shaking as the three helicopter blades picked up speed. The engine and blades sounded different inside the helicopter than they had when I stood outside watching the helicopter start up a few days before. I put my headset on and listened to the aircraft and ambulance radios through my ear pieces. I was familiar with some of the back and forth chatter between aircraft and controllers thanks to years of listening to my scanner back in New York and Dallas.

Once we were all on board and set, Toby got permission to take off. The sequence and sensations of taking off were unlike what I was used to in a plane. When the rotors were spinning at full speed, Toby pulled up gently on the collective, the control on his left side, and we lifted straight up in the air a few metres. Once the helicopter was above the tug, attached to the trailer in front of us, he dipped the nose of the helicopter forward. As we built up speed across the tarmac he glanced over his right shoulder to make sure it was clear and then made a quick 180-degree turn to the right heading towards the blackness of Cook Strait.

I was loving every second of it as we gained speed and altitude. One thing I quickly noticed was the difference between the scene ahead and the one to our right. In front of us was a black sky with a multitude of stars, and it was difficult to distinguish between the sky and the sea below as they merged at the horizon in the distance. To our right were the lights of Lyall Bay and, as we ascended, the lights of the other suburbs and buildings in Wellington's CBD came into view. I suddenly understood how easy it would be to get disorientated flying into darkness without any reference to the ground.

Our training mission was straightforward. We were heading to Taputeranga Island at the south end of Island Bay, which we would circle, and practise using the Nitesun, used by police, TV news and rescue helicopters around the world to light up things on the ground — anything from a person to a car to a football field.

The searchlight is externally mounted and controlled using a remote control from inside the cabin. This remote allows the light beam to be directed in a very wide arc in front of, below and on either side of the helicopter.

When we arrived over the island, I was told to turn on the Nitesun. Once it was lit, we circled the island and I was identifying rocks, trees and any other object I could see. After a few minutes of flying around, we moved the training session further along the coast towards Red Rocks, an area with no houses or people to disturb.

As we continued our training near Red Rocks, I noticed that the helicopter was making tighter and tighter circles and John instructed me to consult the map to identify exactly where we were. As I stared at the map, using a small torch to light it up, I noticed that the helicopter was bouncing around more. I pointed at the map to show John where I thought we were, then continued working with the Nitesun to identify things outside the helicopter.

When we finished, we flew high enough that the lights of Wellington came into view. It was a stunning sight with the buildings of the CBD, the roadside lights circling Wellington Harbour and the lights of the Hutt Valley heading north. The moonlight silhouetting the hills around the region and the night sky were breathtaking.

Toby called the control tower and asked if we could return to the airport via the City. We flew over Brooklyn and John pointed out Wellington Hospital off to our right. We came in over the CBD and Toby circled around as they described what we were looking at.

If for any reason this was my last flight aboard the rescue helicopter, it was going to be a memorable one.

After we landed, John pushed the helicopter trailer back, I helped with refuelling and then watched the blades as John pushed it into the hangar. When it was inside, John pointed to a hole in the wall where a helicopter blade had pushed through at some point. No one was sure when it had happened, he said, as no one had ever admitted to causing the damage. It would have taken reasonable force to create the hole and that would have been enough to damage the blade, damage which might not show up until the blade broke apart on a subsequent flight.

John made it very clear to me right there and then — if I ever caused any damage, the first thing I had to do was admit it and make sure the aircraft got checked. No exception. Not reporting an incident which damaged the aircraft could kill many people and was not acceptable.

Once we were back inside, we sat down with a cup of coffee and had a debrief of the training mission. I had performed well with the Nitesun and even better with the map reading. The helicopter bouncing around while I read the map was actually a test to see if I would get airsick; reading a map in turbulence was a requirement of the job and if I had got airsick I wouldn't have had any more training.

We finished up and John said he would see me on the weekend. I was ecstatic — I was still in.

Once that first flight was done and dusted I started to fly more regularly, usually at least one training flight each weekend.

I was being taught many different skills, but the next big thing on the list was becoming a dope on the rope.

ELEVEN
DOPE ON THE ROPE

USING OUR RESCUE hoist, or winching, is one of the most challenging and rewarding tasks we perform, with everyone on the team — pilot, crewperson and rescuer — working together to ensure a safe operation.

The winch is used for rescuing people from areas where the helicopter is unable to safely land such as in the bush, off a yacht, boat or ship, or directly from the sea. It requires incredible flying skills from the pilot and exceptional teamwork between the pilot and the crewperson operating the winch.

As the person hanging on the end of the wire, you must have implicit trust in the pilot and crewperson above, as well as the equipment being used. The day you don't trust the team or the equipment is the day you don't get on the hook!

The Guardian, a 2006 movie about the US Coast Guard, had the perfect name for the rescuer hanging on the hook. The rescue swimmer who jumped out of the helicopter into the sea was nicknamed 'the dope on the rope'. It's an accurate description because as the rescuer you have little control of your own destiny at the end of the wire. You hang there, look where you are going, and hope like hell that the people above do their jobs well.

The rescue hoist itself consists of a motor spinning a drum which has anywhere from 30 to 76 metres (100 to 250 feet) of steel cable wrapped around it. It is attached to the outside of the helicopter and operated remotely with a control by the crewperson acting as the winch operator. They control the movement of the steel cable and talk the pilot into position directly over the winch area.

My initial winch training was as the rescuer, not the winch operator. The rescuer can either be winched out of the helicopter to the ground or onto a vessel, or for a water rescue, jump directly into the sea.

I was put through a considerable amount of ground training before the time finally came for my first winch training flight. We were going to a flat bit of land on the south coast; we'd get overhead and then I would be winched out of, and into, the helicopter a couple of times.

We headed out over Kilbirnie towards the Karori Light area. I was quite amused as we flew over Brooklyn because I could see my laundry hanging on the line in the backyard of my house. We descended towards the coast and I could see the navigation light sitting in turbulent waters near the head of the harbour. When we arrived, John did a last safety check of my harness.

I listened intently to John and Toby's conversation — the standard winch patter used between the pilot and crew. The crew gets permission from the pilot before opening doors or operating the winch. John opened the left-hand door and I felt the cool air rush into the cabin. He winched out the cable so he could pass me the hook which I attached to the carabiner on my harness.

Once I was safely attached, I removed my seatbelt and my headset and when John gave me a thumbs-up I made my way towards the open door. Seconds later, I was sitting in the doorway of a flying helicopter, my legs hanging over the side, the wind blowing in my face, attached to a steel cable slimmer than a drinking straw. Luckily, I have never feared heights because, if I did, I would have scrambled back into the helicopter right then. We were about 10 metres above the ground, roughly the height of a three-storey building, and I couldn't wait to move outside the helicopter!

I could see John talking to Toby via his headset but I couldn't hear what was being said. A few seconds later John gave me the signal to stand up on the skid and then he pulled a lever which released the small arm that the hoist was attached to. He pushed on the arm and it locked into position 90 degrees from the helicopter.

With that, we exchanged a thumbs-up and a smile and he began lowering me to the ground. As the cable slowly played out I did as I was trained — I pushed off from the skid with my feet until my butt was below the skid level, and then dropped my legs and used my arms to push off the skid. I felt like I was flying.

This was like a slow-motion replay of the only time I jumped out of an aeroplane. Back in Texas, I had done a dual parachute jump where I was attached to an instructor. We stepped out of the cabin onto a small step and then on the count of three I let go of a handle, the instructor pushed off and we plummeted towards the ground. I loved every second of it. Now instead of plummeting at high speed, I was lowered at a more sedate pace — 15 metres a minute — so it took almost 45 seconds to reach the ground.

Dave hanging off the bottom of the winch.

Dope on the Rope

I continually looked down as I was lowered, making sure I was aware of where I was going to land and ready to push off something like a rock or tree branch, if needed. As my feet touched the ground, I removed the safety pin, opened the hook, removed my carabiner from it and held it out to my side. A second later, the hook started moving back up towards the helicopter and shortly afterwards the helicopter flew off. I was standing in a paddock, cowpats all around me, with the biggest smile I have ever had. Being winched out of a helicopter was the best thing ever!

The helicopter returned and then the hook slowly made its way back down until it was sitting directly in front of me. I attached my carabiner and gave a thumbs-up — moments later my feet left the ground and I was on my way back up to the helicopter. I knew I had to keep an eye upwards so that my head didn't hit the skid, but as I made my way back, I took in the surrounding scenery. We were overhead a hilltop, and beyond the cliff was a long drop down to the coast of Cook Strait. It was a clear day and I could easily see the Inland and Seaward Kaikoura Ranges on the South Island. I was surrounded by stunning natural beauty as I was winched up to a helicopter — this was better than anything I had ever imagined.

As I approached the skid, I put my hand up to ensure I didn't hit it. John brought me up to the skid, put his mouth close to my ear and said we would do it again. Soon I was on my way back down. When I came back up John signalled for me to move inside the helicopter. The training for the day was complete.

We had a good debrief and we were all happy that I was progressing well. I felt like I had passed another milestone. They told me that if the weather was good the next day we would do some wet winching — I would jump out of the helicopter into Wellington Harbour.

I woke up to a beautiful day and was excited about what awaited. At base, John talked me through the procedure. Basically, it was like jumping off a high diving board but I had to be sure that I landed feet first — if I landed incorrectly I could break my back! That was incentive enough for me.

A couple of hours later, I had my harness on over my wetsuit and we were hovering over the north end of Ward Island. I put on my mask, made my way to the door and looked down — it seemed waaayyy too high for my liking. John gave me a thumbs-up and patted my back, my signal to jump, but I just shook my head — I thought they were joking!

Eventually, John handed me a headset and Toby told me that we were now about 6 metres above the harbour — for every minute I didn't jump he'd go up by at least a metre and we weren't going home till I jumped. Honestly, I thought we were much higher than that, but regardless I didn't think he was serious. Sure enough, a short time later, up we went. What the hell, if I didn't do it I might lose the chance to ever do it again. I looked at John, he gave me a smile and I stepped off the skid from about 9 metres, almost as high as a tall diving platform at the pool.

I found the jump exhilarating; all the fear was replaced by adrenaline-filled excitement. I hit the water, went under for a couple of seconds, and then came up and gave the okay signal. A minute later, the winch hook arrived and I was pulled up to the helicopter, more pumped up and excited than I had been for a long time. I did two more jumps that day and loved each one more than the last.

Suited up and ready for a wet winch demo.

AS I HAD recently qualified as a rescue diver, my dive instructor suggested that maybe we could have a joint training operation so I could practise my diving and helicopter skills. A couple of weeks later we held the exercise and I learned two very important lessons.

The helicopter landed at Scorching Bay, a small beach near the airport. Dave and John removed both doors from the left of the helicopter to sit in the doorway during the exercise. After a briefing, we got underway. Unlike my first wet winching practice, this time I was wearing my diving fins. Everything was going along swimmingly, as they say, until I started to spin as I was winched up to the helicopter.

I had been told that if this happened, I was to put my arms and legs out like a starfish so my limbs would create drag and slow down the spin. However, instead of slowing down, I went faster. Unknown

Tim Douglas-Clifford and Ruth Zeinert winch a police dive squad member during a training exercise.

78　Emergency Response

to me, the fins acted as little propellers, causing me to spin faster and faster. When I reached the skid, I grabbed it to stop the spin. A couple of revolutions later, the spin stopped and I was brought up onto the skid. All three of the guys were laughing hysterically, the bastards. My first lesson from the exercise was that the 'star fish' doesn't work when you are wearing fins!

THE NEXT MORNING, I had an appointment with New Zealand Immigration as the first step towards my permanent residency. On my way, I passed a dairy and was surprised to see a picture of the Westpac Helicopter, with me hanging below it, on the front page of *The Dominion*, Wellington's morning newspaper.

I was called into a small office where I met the woman looking after my application. She had a folder of paperwork with her and a copy of the newspaper.

We briefly chatted, then she pulled out the newspaper and asked if that was me. When I said yes, she asked how I came to be there and I told her I was a volunteer on the rescue helicopter. It was then I learned my second lesson — how well respected the late Peter Button and the Westpac Rescue Helicopter were in Wellington. She asked me to ensure that the paperwork I submitted included a letter from Life Flight because being part of such an important community organisation would definitely help my application.

At the end of the meeting I was told that it would take up to six months to have my residency application approved or declined. I'm not sure if it was my involvement with the helicopter or not, but ten weeks later my application was approved!

WITHOUT DOUBT, MY favourite place to be is suspended below the helicopter on the end of the winch. This is one of the few places in the world where I find I am living 'in the moment', taking in what is around me, with no thoughts of my to-do list, what I might have for dinner or bills that are due to be paid.

I loved being the dope on the rope, more than anything else I ever did on the helicopter!

Between 1991 and 2010, I spent a lot of time acting as rescuer at the end of the wire. From 2010, once we had a paramedic permanently assigned to the helicopter, they were always available to go down for the rescue, so a favourite part of my job was lost forever.

Aaron Hartle, Nigel Stevens, Hernan Holliday and Dave Chittenden – the WFA paramedics assigned to the helicopter in 2010. *Julian Burn*

TWELVE
LEARNING TO BE A WINCH OPERATOR

I'VE OFTEN DESCRIBED winching as the ultimate team sport. In cricket, two players can score a duck (get out without scoring any runs) and another player can score a century (100 runs) and the team still wins. If anyone involved in the winch operation — the pilot, crewperson or rescuer — gets a metaphorical duck, the chances are one or all of us is going to get seriously hurt or killed.

If my favourite place in the world was hanging below the helicopter, my second favourite place was standing on the skid outside of a moving helicopter.

The next phase of my training was qualifying as a winch operator. Winching is a bit of a fine dance between the pilot and the crewperson. It demands excellent piloting skills, hovering in place, using limited reference points and small movements to correctly position the helicopter.

One of the pilot's biggest challenges during a winch operation is that they can't see beneath the aircraft, the people on the ground or where the winch hook is. They must rely on their crewperson, and the points of reference that they can see from the front and side windows, to move the helicopter where it needs to be.

Meantime, the crewperson acting as the winch operator must work three dimensionally; watching and relaying to the pilot what is happening below, to the side and to the back of the helicopter, as well as monitoring and controlling the amount of winch cable being played out or in. The skill lies in creating an accurate mental picture so the pilot knows where they are in relation to the person being rescued.

To do all this the crewperson must be extremely focused while maintaining a wide view of the situation and able to talk and convey information quickly and accurately. The worst thing for a pilot to hear during a winch operation is silence because this usually means that things are so bad below that the crewperson can't find words to describe it!

There are four types of winch operations that are generally carried out: land winching, where someone needs rescuing from the bush or a cliff; wet winching, where someone needs rescuing from water; or boat winching, from off a boat, yacht or another type of sea-going vessel. The fourth — night winching — is any of the previous three but performed after dark.

The more reference points a pilot can see, such as trees or rocks, the easier it is for them to maintain a steady hover. Good reference points allow the pilot to set a solid starting point so when the crewperson asks for the helicopter to move, they have something tangible to move from.

Wet winching is vastly more difficult as all the reference points, such as waves, are moving, making it nearly impossible to hold a steady hover or to be precise with movements.

During a boat winch, depending on the size and type of vessel, the pilot might be able to find a small reference point, such as the mast, or part of the deck. As soon as the reference is lost, everyone's job becomes more difficult.

The fewer references a pilot has, the more they rely on the crewperson to give them a continuous and accurate description of how the helicopter needs to move. If the pilot cannot see the person being rescued and they do not have a good reference point, it is as if they are flying blind.

One of the ways I first learned winch-operating skills was through an exercise where I sat in a car passenger seat alongside BigD, who was driving with a blindfold on. I had to talk him around the airport tarmac using only our winch patter and varying tone and rhythm. At first it was almost impossible to get him to even go in a straight line, but eventually I could talk him all the way around a track.

My first try at land-based winching didn't go as smoothly as I thought it would, and it was hard not to get discouraged and frustrated. However, looking back, it taught me the power of perseverance. After a few goes it got easier, and with practice I started to get the hang of it.

Operating a helicopter is expensive and every hour spent flying costs thousands of dollars, so, after the first few sessions John and Toby had to decide whether they thought I could master the skill in a reasonable time. I did well enough that they let me keep going at it.

Once I could winch a person at the end of the hook, we moved on to the winch stretcher, which was more complicated again. As I became competent at land winching we moved on to wet and then boat winching. Each stage felt like returning to step one, and was as disheartening and frustrating as the first time at land winching.

My first 'real' winch rescue in 1991 was in front of a large audience of emergency service workers and bystanders, including a TV news camera. A car had gone over a bank in a Wellington suburb and the helicopter was called in to winch a patient from the accident scene. The rescue didn't go perfectly as the stretcher started spinning after I forgot to use a rope that would stop that from happening. We all want things to be perfect the first time we do them, but in some ways mucking up gets that over and done with. A bit like the first scratch on your new car.

In the end, the pilot, Spence Putwain, and I successfully winched a seriously injured person from an inaccessible position and got them to hospital. The rescue featured on the nightly TV news and BigD had videotaped it to use the clip as a learning opportunity for me — or more to the point, he used it to wind me up about all the things I could have done better!

After that winch operation, I performed several more in quick succession.

Winching can be a difficult task, and requires a lot of training and practice to stay proficient and maintain 'winch currency'. Each air operator has their own rules and Life Flight required that everyone involved in any aspect of a winch operation needed to complete some type of winching, either training or real missions, at least once every 90 days. We also had to perform our specialist winching — wet, boat or night winching — at least once every 180 days.

Because winching is such a dynamic activity, many things can affect how well a winch operation goes and how risky it is. Weather, time of day and the currency of the pilot or winch-op can all make it more difficult. In any winch operation, you have one or more people swinging below a moving helicopter and any mechanical problem, with the helicopter or the winch, could have disastrous results.

One lesson that was drummed into me, and that I vividly remembered until my last day on the helicopter, is that the winch does not know whether we are performing a real rescue or a training mission. Training missions have the same risks as real missions and we treated every operation with the respect and diligence it deserves.

ONCE MY WINCH training was successfully completed I was officially accepted as a volunteer.

I finished my contract with Unisys at the end of 1991 and set up a company, Emergency Preparedness Services, making emergency plans and teaching people how to be prepared before, during and after a major event, such as a fire or an earthquake. I wanted to be able to earn some money while spending most of my time volunteering on the helicopter.

I eventually managed to create the life I wanted but in hindsight things were not always as straightforward as they seemed. All the people I've met, all the jobs I've had and all the self-development courses I have done over the years brought me to this place.

John Goldswain winching firefighters into the helicopter during an exercise.

Dave getting ready to lower the nappy harness.

Learning to be a Winch Operator **85**

If I hadn't gone to University, hadn't met Foster, hadn't moved to Dallas or hadn't taken the one-year contract, I would have never ended up on the Westpac Rescue Helicopter.

A friend of mine says that 'life is what happens while you are planning something else'. In many ways, he's right.

THE WINCH I cut my teeth on was quite limited because it only had 30 metres of cable and could only carry one person at a time, making it impossible to winch a patient and their rescuer at the same time.

In March 1993 Life Flight became the first rescue helicopter in New Zealand to begin operating a BK117, also known as a BK. The BK offered many advantages over the Squirrel helicopter.

It is a twin-engine helicopter, giving us huge safety margins during winch operations and when flying offshore. The helicopter could carry two stretcher patients, plus two EMT/paramedics in the cabin. It also had a new and better winch.

We now had 76 metres of cable and could lift 272 kilograms, meaning we could winch two people at a time. This winch moved at up to four times the speed of the old one, so we could complete winches faster and more safely.

OVER THE YEARS, I have performed hundreds of winch operations and most have gone well. There have certainly been some learning opportunities along the way and I have taken each of those and been better off for it.

In the end, winch currency, or the lack of it, is what made me decide to resign from the helicopter.

THIRTEEN
MEDIA SPOKESPERSON

'**WE MIGHT AS** well just paint a big fucking "A" on the side of the helicopter,' one of the pilots exclaimed as he read a news story recounting a rescue they had performed the day before. He was frustrated that the article identified 'A helicopter', not 'the Westpac Rescue Helicopter', as the one which had performed the rescue.

That afternoon we were called out to rescue a tramper who broke their ankle in the bush. When we were back at base the phone rang, the same pilot answered the phone, listened for a moment and then told the person on the other end to 'get fucked' and slammed the phone down.

When I enquired who that was, he told me it was a reporter from *The Dominion*. It seemed obvious to me why the paper called us 'a' helicopter if that is the way they were treated on the phone!

My only previous experience with the media was my Dan Rather encounter back in Dallas, but I could see that there was plenty we could do to get better media coverage. However, I was still new and nervous to suggest anything to the pilot.

I decided to park any ideas I might have until I had earned my stripes and felt confident to suggest anything innovative.

After that day, if I answered the phone, I would politely answer the reporter's questions and give them whatever information about a rescue I could. I'd also remind them that the helicopter was called the Westpac Rescue Helicopter and, unsurprisingly, when the article appeared the next day, the proper name was used.

Slowly and surely, as the number of mentions of our proper name appeared, attitudes towards the media improved a bit. The general attitude was that the media 'always gets everything wrong' and they couldn't be trusted. My experience was that if we took the time to work with them and were open and honest, then the reporting was usually accurate.

At first it was a buzz to see my name in a newspaper article, but every time I saw the Westpac Rescue Helicopter name I felt good that things were improving. One day Peter Mairs, our General Manager, commented how well I was doing with the media mentions and that it was making Westpac, our major sponsor, happy.

I thought I was on to a good thing. But I had never heard of the 'tall poppy syndrome' and it turned out that I was wrong!

I was too naïve and stupid to realise that as the media mentions increased, it seemed to me that a lot of people were getting steadily pissed off. They were happy to see the helicopter name correct but suddenly this new American guy was getting all the credit in the media.

MY PROBLEMS COMPOUNDED when I became the 'Eye in the Sky'.

Each morning, if the weather was suitable, our helicopter would fly around the Wellington region for 30 minutes and reporter Paul Brennan would give traffic updates across three radio stations.

After I finished with Unisys, at the end of 1991, I'd go to the hangar every morning and help John reconfigure the helicopter with seats. People in the community would pay $50 each for the opportunity to be on the traffic report flight. It was a great fundraiser for Life Flight.

I often joked with Paul that if he ever gave up doing the reports I wanted to do them. I could think of nothing I would want to do more than fly around Wellington for 30 minutes every morning!

One day he told me that he was going to be reading the morning news reports for a couple of weeks so I better come up with him for a few days and learn the ropes. I was a bit gobsmacked because I didn't really think I'd ever have the opportunity, but suddenly it was mine.

I did a couple of flights with him that week and suddenly the next week I was doing live radio reports on the three most popular radio stations in Wellington. Each of the stations would call the helicopter's cell phone at a specified time and I would give traffic updates. I loved it, and the announcers at the stations were loving my American accent.

After the initial two weeks, I filled in for Paul whenever he could not do them. One day, in early 1992, he told me that he was moving on from doing the reports and they would be mine from then on. I thought this was fantastic.

I found myself part of Lindsay Yeo's morning show on ZB, and part of the Morning Crew, with Polly, Grant and Nick on ZM. My distinctive voice and my unique way of signing off from each report was becoming well known around Wellington.

If there happened to be an article in the paper about a rescue the helicopter had performed, then one of the announcers might ask about it during the broadcast. Peter told me that the fundraising money had increased and he thought it was because of the mentions I was giving the helicopter during the traffic reports.

I would have to admit that I didn't mind the attention the traffic reports were bringing my way. Even though I never mentioned my business, the name and voice recognition was helpful when I did sales calls.

THEN SUDDENLY, MY world came crashing in.

One day Peter Mairs called me into his office and told me that there was a move to get rid of me from the helicopter crew because there was a feeling among some of the pilots and other crew that I was getting too big for my britches. He didn't agree, he saw the value of what I was doing for the Trust, but he didn't interfere with operations — this was my problem to solve and I better solve it quickly.

I felt like I had been taken into an alley and beaten to within an inch of my life. As much as I loved the traffic reports and flying around every day, the main thing I loved and cared about was the rescue work.

After I gathered my thoughts, I asked several of the others to meet with me. I told them that Peter had warned me that I was about to be kicked off the helicopter because of all the media stuff and I didn't want that. I offered to quit traffic reports that day, and to never speak to another reporter. I would do whatever it took to remain a part of the helicopter team.

I must have said enough and been seen to be genuine about it, because they agreed to rethink their decision. I was told I should keep on doing the traffic reports for now.

Once I had convinced them that I would give all the media stuff up to retain my position on the helicopter, Peter convinced them how important all the media had been for fundraising and bringing extra exposure to the Trust.

In the end, I was extremely relieved when the decision was made that I could continue as crew on the rescue helicopter, and I would become the media spokesperson for Life Flight. It was a solution that made me very happy, mainly because I remained on the rescue helicopter.

The media stuff was great, and I appreciated how it helped the Trust, but regardless of what other people thought, I was doing it for the good of the Trust, not for personal gain or notoriety.

The whole experience also made me very aware of how hard I had to work to make sure that other people's names, whether it be the pilot or paramedic, appeared in the media as often, if not more often, than my own.

Within Life Flight, I also learned who was willing to front media interviews, if they were asked nicely and enough times; some of the others would front a TV camera when first asked, some would have to be asked twice, and some wouldn't get in front of a camera unless someone held a gun to their back, forcing them to do it.

Our relationship with the media was mostly strong and I worked hard to ensure that they had good information, and in return they were accurate in their reporting and were sure to use the 'Westpac Rescue Helicopter' name in each story.

OCCASIONALLY, I DID have learning experiences, all of which we used to keep our media policy up to date and relevant.

On a cold winter's day in 1998, we were called to two rescues in a short time. The first call was for a man who had fallen 3 metres from a balcony at his home. We suspected he was suffering from spinal injuries so took standard precautions to ensure that his injuries weren't aggravated during the flight.

Soon after dropping this patient at hospital we were called to a patient who was shot in a domestic incident. Gun violence is not very common in New Zealand so the scene was swarming with police. Once the scene was declared safe, we attended to the patient who was suffering from multiple gunshot wounds. Our paramedic worked quickly to stabilise her enough to transport her to hospital.

I had snapped a couple of photos at each of the scenes and once processed and printed I labelled them 'Balcony 1', 'Balcony 2', 'Shooting 1', etc. and sent them to Mark at the *Evening Post*. Mark was a reporter I knew well and trusted implicitly.

The next day we flew the patient who had fallen from the balcony to the specialist spinal unit at Burwood Hospital in Christchurch. While we were flying, I got a phone call from Mark who had an urgent question regarding the photos from the missions yesterday. He asked me to describe the scene of each incident and I replied that it didn't matter because the filename would tell him which scene was which.

He agreed that my naming was good but still asked me to describe the scenes. Against my better judgement, I asked him why he was asking all these questions. He said that there had been a bit of a mix-up with the first edition of the paper and the wrong photo had been used with the wrong story.

Mark was highly stressed and embarrassed by the mix-up but I wasn't concerned because there were no privacy issues and no damage was done.

This did give me an opportunity to put better processes into place around emailing photos to the media. If we were going to send photos for more than one mission we would send them in separate emails. Of course, this wouldn't help in all cases, because how much more specific could you get than naming a photo something like 'balcony' or 'shooting'!

Soon after, Mark delighted in how easy it is for other professionals to get it wrong, occasionally. I had been on a mission overnight and Mark rang me early the next morning to get more information before his deadline. He apologised profusely for calling so early but it was okay because I was already awake.

During the phone call, I got distracted and he asked what was wrong. I said I had misplaced something, and he asked what that was. 'My cell phone,' I replied.

Mark paused and then asked what I was speaking to him on.

'Oh yeah, thanks, it's like losing your glasses when they are on top of your head,' I said, hastily trying to play it down.

'But your glasses aren't talking to you,' he said.

I didn't give this much more thought until the following Saturday when I read a weekly column called 'The Last Word', which poked fun at Wellingtonians.

'One of Wellington's respected rescue helicopter crew interrupted his early morning phone interview in a burst of panic. "I've got to get to work and I can't find my cell phone anywhere in my room," he wailed as he struggled with the wake-up blues. Shouldn't have been that difficult – he was using it at the time.'

I AM VERY proud that over the years I created a top-class media programme and also set up and produced the content for Life Flight's first web page, Facebook page and Twitter account. I always believed that sharing our stories, and staying in touch with our supporters, via media and then social media, was an excellent way to help raise funds to keep our services going.

Looking back, I still find it very amusing that back in Dallas I had made a life decision to never get in front of a TV camera again. Just goes to show, you never know.

FOURTEEN
TEAMWORK AND CREW RESOURCE MANAGEMENT

LIFE FLIGHT PERFORMS a wide range of missions including responding to the scene of accidents and medical emergencies, carrying out search and rescue missions, both on land and at sea, as well as transferring sick patients between hospitals. Occasionally the helicopter also assists the police or army bomb squad as well as an array of other missions that crop up from time to time.

One of the many things I loved about the job was that I never knew where I would end up or what I would be doing when I turned up at work for a shift. One day might be quiet, with no missions, and the next we'd be run off our feet with one mission after another. It was not a job suitable for anyone who needs certainty or routine in their work life!

Saving lives, whether on the helicopter or the fixed-wing, was always a team effort. We all relied on each other to get the mission completed safely but none of the team would get where we were going without our pilot's amazing efforts.

Teamwork, in an aviation environment, is known as Crew Resource Management (CRM). CRM has many different components, but in simple terms it means that everyone in the aircraft works together to keep each other safe. If anyone feels unsafe, or sees something dangerous, we are taught to speak up immediately. Pilots cannot be looking everywhere at once, so they rely on input from others.

In a sound CRM environment, everyone on the flight crew, whether a pilot, crewman or medical team member, feels confident to speak up, and even to disagree with another person, even if they are more senior in rank. In our operation, the pilot always has the final say on flight decisions, but if any of the members of the team thinks we should abort a mission, the mission will be aborted, if it is safe to do so.

One of the benefits of being well trained in CRM was that team members were able to put personal feelings aside and get the job done. We didn't all love each other or socialise outside of work, but when we got on the helicopter we generally worked well together.

Dave getting ready to winch paramedic Dave Chittenden during winch training. *Derek Quinn/111 Emergency*

I CANNOT BEGIN to describe the admiration I have for the pilots I have flown with. They fly during all hours of the day and night, in testing weather conditions, off mountains, out at sea or wherever else we were asked to respond. They could put the helicopter down in places smaller than a tennis court, or hover in place over a pitching yacht to save an injured sailor.

All our pilots were highly skilled and experienced. Each of them had to meet the strict requirements of the New Zealand Air Ambulance and Air Search and Rescue Standard which included a minimum of 2000 flight hours for the helicopter pilot and fixed-wing captain.

One of the hardest decisions a pilot makes is when to decline a flight due to weather conditions, mechanical issues or any other safety issue. They constantly worked under the real, and perceived, pressure that if we didn't fly, someone might die. However, they

Dave, Jon Leach, Dean Voelkerling, John Goldswain and Ruth Zeinert.

needed to weigh up the safety of everyone in the aircraft based on flight conditions, not the condition of a patient requiring urgent surgery or the dilemma a tramper has managed to create for themselves on a mountainside.

The pilot carries responsibility for any decisions that the team makes. If there is an accident or incident during a flight then the pilot would have to defend these decisions. They can also lose their licence if things go wrong. Of course, the people who review these decisions would do so over a cup of coffee, in a warm dry office, with the benefit of hindsight — they are never reviewing the decisions under the same conditions that the pilot was making them at the time.

Mike Hall, one of the pilots, wrote a list of all the considerations he had to make before he could accept a flight. It was an enormous list which included knowing the weather in Wellington, where we wanted to land, and everywhere in between. The list was extensive and I was stunned by the number of things they had to consider for every flight which I never knew about.

A CREWPERSON'S FIRST role is to help the pilot ensure the safety of everyone on board. The crew could give safety briefings, make sure

Dave, Julian and Harry on the job.

Teamwork and Crew Resource Management

people are secured for take-off and landing, and assist if there is an emergency on board.

Crew look after safely loading and unloading patients and aid the medical teams who are experts at their job, but might be unfamiliar with where some of the aircraft equipment is stored or how to use a model of machine they are unfamiliar with. All fixed-wing crew have some level of medical training, so that they can help with CPR or other medical emergencies in-flight.

Helicopter crew have added responsibilities including operating specialised equipment such as the winch or the Nitesun, and are trained in special skills, such as aerial observation to help locate people missing at sea or in the bush. They are also trained to at least an EMT level so that they can assist the paramedics with patient care, and, at the scene of an accident with multiple patients, can look after some of the injured.

On the ground, a crewperson answers the emergency phone lines, works closely with the hospital teams to coordinate patient flights and organises ambulances to meet the fixed-wing to transfer patients to or from hospital. All crew were also experts at completing paperwork, cleaning and checking equipment and dealing with myriad other things that might occur on base.

Part of the 2003 team: John Goldswain, Jon Leach, Phil Harris, Grant Withers, Steve Reeve (back row) Ruth Zeinert, Brian Taylor, Dave, Dean Voelkerling (front).

A Life Flight crewperson is somewhat like the conductor of a very sophisticated orchestra. There are many people and organisations required to get the aircraft into the air and Life Flight is the organisation that pulls them all together to play the same tune.

TREATING A PATIENT in the back of an aircraft can be difficult and challenging for the doctors, nurses and paramedics. They have to contend with turbulence, the confined space in the back of the aircraft, difficulty of accessing parts of a patient's body, and have to have special training to deal with the effects altitude can have on a patient. A patient who is stable on the ground can suddenly deteriorate because of the changing air pressure as an aircraft gains altitude.

Colin Larsen, Dave, Keith Frewen, Logan Taylor and Alan Deale, 2009.

They also need the appropriate expertise and qualifications to work on their own. In the hospital or at an accident scene, there are many people close by to assist when required. In the back of an aircraft, the only other person they can turn to for help is a crewperson.

IN 2010, WELLINGTON Free Ambulance and Life Flight agreed to roster a winch-trained paramedic onto the helicopter full-time. This eliminated many of the problems we had where a paramedic working on a road ambulance would be delayed responding to a helicopter mission. Having the paramedics on base meant we could use them on more than just ambulance missions. We now had a permanent winch-trained 'dope on the rope' and an extra pair of experienced eyes for our search and rescue missions.

When the paramedic had a seriously injured or ill patient in front of them, they had to balance the amount of time they remained at

The ICU flight nurse team with pilot Alan Deale, 2009.

the scene, trying to stabilise the patient, against the importance of getting the patient to hospital. In the ambulance world, this is known as the difference between 'loading and going' or 'staying and playing'.

Transporting a patient without a stable airway can be deadly. The helicopter paramedics began performing a skill called RSI (rapid sequence intubation). RSI involves giving a patient drugs to sedate and paralyse them, allowing the paramedic to insert a breathing tube down their throat. This is the same procedure that doctors perform in the ER or an operating theatre, but the paramedic is doing it in the back of the helicopter or ambulance.

Once the paramedic starts giving the drugs there is no going back. As soon as the patient is paralysed they can no longer breathe on their own; the correct insertion of the breathing tube is critical. After the tube is inserted then a mechanical ventilator, or a manual bag mask, is used to force oxygen into the patient's lungs.

As the paramedic's training and skills increase, so does the amount of teamwork required, because they need assistance from the crew or pilot in order to carry out many procedures.

FIFTEEN
SEARCH AND RESCUE

MANY OF THE most exciting and rewarding missions I was involved in were Search and Rescue (SAR) missions. Many times, our team had the satisfaction of knowing that if we had not been there, a person would not have survived.

We would receive our SAR missions from either the New Zealand Police, who coordinated most searches on land and close to shore, or the Rescue Coordination Centre New Zealand (RCCNZ) who coordinated searches for missing aircraft or when an Emergency Locator Signal was detected by satellites circling Earth.

No matter who sent us out, the required outcome was the same; locate and rescue people in distress.

SAR is truly a team effort. A helicopter is an amazing tool for helping search large areas quickly, but it is useless for searching under trees in thick bush. I can recall several searches where a person was found as soon as the noise of the helicopter rotor blades and the disruption of the downdraft left the area, and ground searchers could hear a person calling for help.

The Police or RCCNZ use all available information to determine a search area and then send appropriate resources to carry out the search. We worked closely with the Police SAR squad for land searches and the Police Maritime Unit for marine searches.

We also worked alongside the Wellington Airport Marine Unit, Coastguard units, Surf Life Saving and LandSAR volunteers, depending on the location and type of the search.

Search and rescue in New Zealand is highly reliant on volunteers and I have great respect for the many people who spent hours and days tramping through the bush while we flew around above them.

I was involved in over 450 SAR missions during my career and no two were the same. The weather, the time of day, the angle of the sun, or something as simple as the colour of the clothes a missing person is wearing could be the difference between us spotting or missing them.

Many of the searches were for missing scuba-divers along Wellington's coastline. Unfortunately, most often, divers failed to use the buddy system (sticking with another diver while underwater) so they couldn't find each other when they surfaced. We would start searching from the diver's last known location, and then expand the search from there. Wellington coast's strong tides can sweep a person many kilometres in a short space of time.

Helicopters and boats would scour the search area, looking for the missing diver. In one of the Wellington region's longest searches, a diver was discovered after being lost at sea for three nights. Our helicopter was used for the first two days of the search, but we were stood down when it seemed the diver was lost. Miraculously, he was discovered alive by a team aboard the police launch *Lady Liz III* just as they abandoned their search and were headed home.

Unfortunately, not all the divers we searched for were that lucky. Many were discovered by the Police Dive Squad at the bottom of the sea floor, and others never found at all.

A DIRECTIONAL FINDING (DF) unit is used during a beacon search to track a beacon signal. A display in the cockpit points us in the direction where a signal is being emitted from. The search for a beacon is often complicated by the signal bouncing off surrounding mountains or being concealed by an overturned hull, or in aircraft, by its Emergency Locator Transmitter aerial ripping off during a crash landing.

As we approached the area where the signal was tracked to, we would have to revert to aerial searching to spot the person in distress. Few things are as satisfying as spotting a person frantically waving at us, thankful that help had arrived.

Many of the SAR missions I recount in this book, the Sounds Air crash, Terminator and the Anzac Day crash all involved the use of an emergency beacon.

Unfortunately, many beacon activations were false alarms. Beacons were often accidentally set off when their switch was knocked on when the beacon was thrown into a bag or drawer. These searches were costly for RCCNZ, but every signal detected must be resolved, no matter what.

One of the most amusing beacon searches I was involved in was when a signal was tracked to the rubbish tip near Porirua. Brian landed the helicopter near where we thought the beacon was located, and then he, John and I used a handheld radio to track the signal. We narrowed the search to an area of the tip and then got the city council to come out and dig it up. While we were waiting for a council worker to arrive the three of us amused ourselves by finding hidden treasures amongst the trash. In the end, the beacon was dug out of the tip; it had been thrown out and had its switch knocked on.

John Goldswain and Brian Taylor with some of the treasures found in the tip.

In the late 1990s, a new, more sophisticated Emergency Locator Beacon and satellite tracking system was introduced worldwide. GPS-equipped beacons could now send the GPS location of a person in distress, reducing a lot of the search time from an SAR, resulting in quicker rescues.

THE SEARCH WAS only half the mission. When we set off on a beacon search we were always prepared to perform a winch rescue, because experience told us that most beacons were set off in a remote location where it would be impossible to land.

Thousands and thousands of people owe their life to the fact that they had the common sense to carry a portable location beacon (PLB) when they were going into remote areas.

SIXTEEN
NO ROOM TO RESCUE THEM ALL

AFTER DELIVERY OF the BK117 helicopter in 1993, our team routinely carried out rescues 200 to 300 kilometres out to sea. We also performed missions beyond the Chatham Islands, a small group of islands more than 700 kilometres to the east of Wellington. That's a lot of time and distance to spend over water, particularly in a helicopter.

When we travelled offshore, we had to wear an immersion suit: a big, bulky and heavily insulated suit. The suits are hot and uncomfortable, but also give the wearer the best chance of surviving should we crash and end up in the water.

The biggest limitation to heading offshore was the limited amount of fuel that the BK117 could carry. At best, the helicopter can travel about two and a half hours, or 500 kilometres on a tank of fuel. This doesn't account for the fuel reserves required on any flight; the pilots always aim to land at their destination with plenty of fuel in the tank.

Soon after the BK arrived in Wellington, we worked with our helicopter engineers to create ways to increase the amount of fuel we could carry. Our initial solution was to secure as many 210-litre drums of fuel as required in the cabin. Each drum increased flight time by about 40 minutes. We used an electric pump to refuel the main tank from a drum, with a manual pump as a backup.

EARLY IN THE evening of 16 October 1993, just months after we moved to our new BK117 helicopter, an emergency locator beacon activation was detected around 260 kilometres southeast of Wellington. Sunset was at 7.45pm, only a couple of hours away, so the rescue coordination centre requested that we send both a plane and helicopter to the area.

Toby was the helicopter pilot and John and I were on board. We loaded several drums of fuel and set off for the hour and a half flight to the area. As it was my first major offshore trip, I was feeling a little nervous. We had all our emergency gear on board but if anything went wrong we were on our own; no other helicopter in the region could fly as far out to sea as we could.

Tom Sunnex was flying the plane and Phil Harris was with him to operate the directional finding unit. Since the plane is nearly twice as fast as the helicopter it arrived in the search area long before we did.

About an hour into the flight, Tom informed us that they had spotted a life raft floating in the sea. I had only winched off a life raft once before in training so John reminded me about the

Pilot Grant Withers and Dave getting ready to head offshore wearing immersion suits.

techniques I should use. He said that he was sure he wouldn't need to, but if I had any problems he could move to the back and take over the winching.

En route we pumped one of the drums of fuel into the main tank to make sure we had enough fuel to carry out the rescue. I could pump the other two drums on the way home.

I was wearing a winch harness so, if required, John could winch me down to the raft. When we arrived overhead the life raft it was just starting to get dark and the raft was jumping around in large seas. This was going to be my most difficult winch yet and I focused on getting it right.

I was as stressed as I had ever been on the helicopter but easily managed to get the two rescue harnesses to the raft. The people put them on and within a few minutes of us arriving we had them on the way up to the helicopter. I was glad we had finished the rescue before it got completely dark.

As I watched them come up towards us, two more heads popped out of the opening in the life raft. I called out on the intercom that there were 'two more' and Toby replied: 'Two more what??'

We had plenty of fuel but now we were looking at winching two people off a life raft in the dark.

When the first two were safely in the helicopter they told me that there were still four people on the raft. This was a major problem. With three fuel drums in the cabin, we didn't have space for me and six others.

I was told to dump the empty fuel drum into the sea. While I was doing this, the guys up front worked out how much fuel we had in the tanks and how much we needed, and then told me to throw one of the full drums into the sea also.

Ha, easy for them to say! A full 210-litre drum of fuel weighs about 170 kilograms and is a challenge to manhandle at the best of times, let alone in a cramped helicopter cabin. With the help of one of the rescued fishermen I got the drum to the door and then pushed it out. It struck the skid on the way down, sending a shudder through the helicopter.

Now that we had room, we still had to rescue the other four people. I put a glow stick on the winch hook which helped me keep track of it as it headed towards the surface of the sea. Luckily, the people in the life raft also shone a torch. Over the next 20 minutes we managed to rescue the remaining four and then headed home.

We ended up rescuing five men and one woman who had been working on a commercial fishing boat. While pulling in a net full of fish, they were broadsided by a large wave which tipped them over. The crew managed to grab their emergency beacon and get the life raft off the deck before the ship sank.

On the way back to Wellington, I had to work my way around the six of them to pump the last of our reserve fuel into the helicopter.

WE WERE PLEASED that the drum fuel system worked so well. Shortly after this flight our helicopter engineers delivered a 300-litre fuel tank which could be secured in the cabin of the BK. The 300 litres of fuel gave us an extra hour of flying.

If we required more than 300 litres we could still carry drums of fuel which could then be pumped into this auxiliary tank.

This 300-litre tank was still being used when I finished flying in 2016 and helped us perform many extended range missions.

SEVENTEEN
HOSPITAL TRANSFERS — ICU

THE MISSIONS THAT I tend to remember, and that got the most media attention, were the daring and adventurous ones. However, the missions that have made a difference to the most people's lives are the inter-hospital transfers.

With such a small population spread across such a large area, it is not practical to have hospitals or acute medical services available in every town or city. Instead, New Zealand has a network of clinics, medical centres and small hospitals which look after the majority of the patients in their local communities. Patients who cannot be cared for locally must be transferred to a larger regional hospital or, in the case of acutely ill patients, to one of the tertiary hospitals located in Auckland, Hamilton, Wellington, Christchurch or Dunedin, for the highest levels of care.

In 2014 when I finished as Operations Manager, hospital transfers made up about 50 per cent of helicopter missions, but 100 per cent of the fixed-wing air ambulance missions carried out in Wellington each year. In 1991, the fixed-wing was only used occasionally and carried about 80 patients. By 2013, the Wellington-based fixed-wing transported about 700 patients, more than double the number carried on the helicopter. It is little wonder we often referred to the fixed-wing air ambulance as our 'silent hero'.

Specially trained flight doctors and nurses from Wellington

Hospital provide the medical care for sick children and adults during these inter-hospital flights. A midwife from the hospital joins the team when a pregnant woman is being transferred.

From 2005 through 2015, Life Flight also ran a fixed-wing air ambulance based in Auckland. In addition to transferring ICU and NICU (Neonatal Intensive Care Unit) patients, the Auckland team worked closely with PICU (Paediatric Intensive Care Unit) to transfer sick children from all over New Zealand to Starship Children's Hospital. The Auckland plane was also the only one able to carry out a complex transfer of ECMO patients, where a specialist team transported a patient while their heart and lung functions were performed by a machine.

When performing an inter-hospital transfer, the aircraft becomes, in essence, a mini ICU. The ICU stretcher has a ventilator to keep the patient breathing, a vital sign monitor to keep track of everything from their heart rate to the amount of CO_2 in their system, a defibrillator in case their heart needs to be restarted and several syringe pumps which can pump much-needed medications into their bodies.

The patient may have been transported because of traumatic injuries received in an accident or because they have an unknown or complex medical condition which is slowly shutting down their body functions.

Regardless of why they are transported, the medical team needs a broad range of skills and knowledge to care for the patient during transport. Most of the flight nurses are very experienced and have often worked in ICU before the registrar with them even entered medical school. The smartest registrars are the ones who recognise that when their flight nurse 'suggests' something, they are not really suggesting, they are telling the doctor what needs to happen!

Many of the ICU transfers are as urgent as an ambulance flight from the community. When a patient is suffering from a serious condition, such as an abdominal aortic aneurysm or blocked arteries requiring a stent, a quick transfer for urgent intervention is required to save their lives.

Once a patient is well enough, they are returned to their local hospital to continue their recovery. Bringing them closer to home means that they are closer to their family and friends as well as freeing up a bed in the tertiary hospital.

Whether on the helicopter or the fixed-wing, a crewperson takes an active role in assisting the medical team during an ICU transfer. The crew has the training and experience to set up the medical equipment or assist with transferring spinal patients onto our special gear.

The crew on the fixed-wing must also coordinate road ambulances to transport their patient from the hospital to the airport and, once at the other end, from the airport to the hospital.

The crew, pilots and medical staff must be flexible. Even though their shift is scheduled to end at 6pm, they might find themselves midway between Invercargill and Wellington, or stuck in Auckland traffic at shift changeover time.

The fixed-wing pilots, crew and medical teams got very little public attention, (before the days of reality TV) because ICU and NICU

The Metro fixed-wing air ambulance operated between 2000 and 2012.

transfers were not seen as sexy or glamorous, but they truly were the people who worked the longest hours, under all types of pressure, day after day. The fixed-wing could only be our 'silent hero' because of the group of silent heroes that work on it.

When I started at Life Flight we used any available plane for a fixed-wing ambulance job. As the service got busier we moved to a Golden Eagle 421, then on to a Piper Cheyenne PA-31T. In 2000, we started flying in a Metroliner SA227, which was capable of carrying two stretcher patients and two medical teams. The Metro was replaced by a Jetstream J32 in 2012.

John Goldswain developed the stretcher system which was used in the fixed-wing planes, as well as the stretchers which carried the patient and all of the specialised equipment.

We worked closely with Wellington Hospital on all aspects of the transfers. Over the years, I had the privilege of working with so many talented nurses and doctors. I would fail to mention them all by name, so will not even try, but, behind the scenes John and I spent many hours meeting with Delia, Henny, Peter, Alex and Karyn regarding adult transfers, and Rosemary, Vaughn, Sara and Sarah regarding the neonates. Together Life Flight and Wellington Hospital created a world-class air ambulance service.

Dave and ICU doctor Laurence in the back of the helicopter.

EIGHTEEN
HOSPITAL TRANSFERS — NICU

THE SMALLEST BABY I had ever seen lay in the palm of my hand. Her tiny body only extended from the tip of my fingers to my wrist and she weighed about 500 grams (1.1 lbs). I watched in amazement as the doctor and nurse put a breathing tube down her small throat. Her veins were so miniscule that it looked like it would be impossible to start an IV line, yet somehow they managed to start several of them.

When we picked up this critically ill baby, the team first spent several hours stabilising her before we flew the baby to Wellington. She spent four months in NICU (Neonatal Intensive Care Unit) before being discharged back to her home hospital and soon after she finally went home with her parents and then went on to live a healthy life.

She was just one of the nearly 600 premature babies I helped transfer to the NICU, or back home, over the years.

In New Zealand, nearly 5000 babies are born prematurely (before 37 weeks) each year and many arrive as early as 24 weeks. As medical science improves, so does the viability of saving younger and younger babies but they often must spend days or even months in a NICU located at one of the tertiary hospitals.

The advanced care the NICU medical team provides started the moment our team arrived at the baby's side. Our team carried skills, drugs and equipment which are not available in smaller hospitals so could begin more aggressive treatment straight away. Often a baby's condition would begin to improve before we even headed back to Wellington.

As a crewperson, there was nothing medically I could do to assist the medical team during a NICU transfer. However, while they were busy stabilising the baby for transport, I was very good at little things like making excellent cups of tea, or fetching bits of medical gear for them. Stabilising one of these babies could sometimes take hours, so a little TLC for the medical team was always appreciated.

Another thing I did do well was offer emotional support to the family members waiting for their sick baby to be transferred. It's a frightening and stressful experience, especially for the parents, sitting in a room where the temperature is uncomfortably hot, watching strangers poking and prodding at the smallest human being they have ever seen, instead of experiencing the expected joyful birth and return home with a new baby.

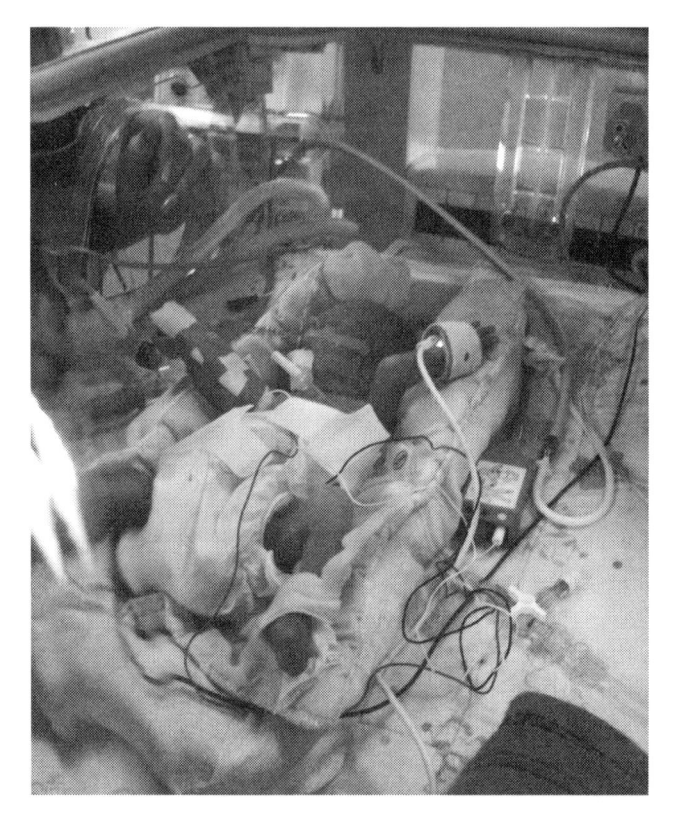

Cameron Sigmund, born weighing only 715 grams, in an incubator.

While the medical team worked, I would explain what the team was doing and what was going to happen from here. Most people were grateful for the practical advice like how to get to Wellington Hospital, where to park when they got there, and where the NICU unit was situated.

As Operations Manager, it was always gratifying to receive a letter from the family of a baby we had transported which specifically thanked the crewperson, often by name, who looked after them so well. If my team was offering so much support that people remembered their name on one of the worst days of the family's life, they were doing damned well.

I also had several friends who had premature babies themselves, or full-term babies that needed NICU care, so I knew personally how tough it was. They all received help from the Neonatal Trust, a charitable trust which gives support to families of premature or sick full-term babies as they make their journey through neonatal care, to the transition home and onwards.

I was so impressed by the difference that the Neonatal Trust makes to neonatal families that I am in the process of becoming a trustee on their board, so I can support their work in a more direct way. You can see more about the good work they are doing at https://www.neonataltrust.org.nz/

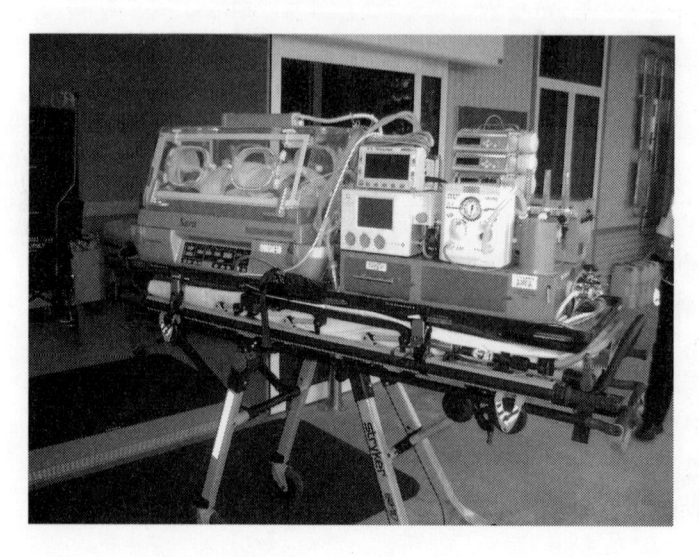

NICU transport incubator on an ambulance stretcher.

NINETEEN
SOUNDS AIR CRASH

LATE ON THE afternoon of 29 January 1996, a small passenger plane operated by a small airline called Sounds Air, left Wellington Airport for a 25-minute hop to Koromiko Airfield near Picton. On board were the pilot and five passengers — they never reached their destination.

The plane disappeared from radar in rugged hill country near Mount Robertson, a peak which lies between Blenheim and Picton within the Robertson Range on the east coast of the upper South Island.

Air traffic control was not getting any response to their radio calls. However, planes in the area started picking up a weak distress signal from an Emergency Locator Transmitter (ELT). It was not looking good.

Around 5pm, we got the call-out and within 15 minutes the helicopter took off to search for the plane, which was now well overdue. Steve Oliver was the pilot, John the crewman, and Dean Voelkerling was the paramedic, with two others, Maurice Lobb, a training crewman, and Pete Collins, the ambulance officer who had been working with Dean on the road ambulance that day.

I wasn't on duty, so I arrived at the base soon after the helicopter had left. During my time in the emergency services, few things frustrated me as much as missing out on a 'big job', and this was looking like a big job. Like everyone else on the team, I wanted to do what I am trained to do. But, although I wasn't on board, I could still help. I took over the emergency phone, ensuring that the helicopter team wasn't disturbed by anything unrelated to the search.

When the helicopter arrived in the search area, they picked up the weak ELT distress signal. The technology in 1996 was not as advanced as today, so although the signal could bring rescuers to the general area, it took some finessing of the electronic gear and a visual search to find the exact location. The team was sure that they were in the right area, but cloud shrouded the mountainside, hampering the effort to locate the crashed plane.

The trick of searching from a helicopter is to have a picture in your mind of what you are searching for. Untrained observers often make the mistake of looking for a white aeroplane with blue stripes, two wings and a propeller; however, that is not what you have to search for. You must look for broken branches, disturbed bush, bits of wreckage in the tree tops, or anything else that might look out of place.

The team narrowed down the electronic search to a specific area. However, the cloud was impeding the process so they had to wait for it to lift before the visual search could begin. Fortunately, as it was summer, there was still plenty of daylight left.

As soon as the cloud cleared the visual search began. Pete spotted what he thought to be the wreckage through some broken branches in the canopy and called the target to the others. The only way to confirm if this was the crash site was to winch Dean down. They spotted a small hole in the trees and lowered him to the hilltop below.

As he descended, he spotted the mangled wreckage. The plane was so smashed up that he doubted anyone could have survived. He radioed back to the helicopter to let them know this was the location of the crashed plane, and advised he was going to check for survivors, although he had little hope of finding anyone alive.

The helicopter flew a short distance away so that the down draft and noise wouldn't distract him.

When he reached the plane, he saw people still inside and began to check to see if anyone was alive.

He was stunned to find the pilot alive, holding one of the passengers and trying to care for her. Clearly in shock, he told Dean that 'she had just stopped breathing'. Dean checked her and confirmed she was dead and there was nothing more they could do for her.

After checking the others, Dean called the helicopter to report that there was one survivor and five dead.

Although the pilot was badly injured, with Dean's help, he managed to extricate himself from the cockpit. With assistance, he staggered the 20-odd metres to the winching area. Dean got him into the rescue harness and sent him up to the helicopter. Dean remained on the ground so he could double-check the other five people for any signs of life.

Pete had not heard the whole radio message, and although he knew there was a survivor, he didn't expect him to come up on the winch alone and wondered where the hell Dean was.

Pete began treatment and a few minutes later, once Dean was winched back in, they headed back to Wellington Hospital with the critically injured pilot.

While they were en route to Wellington, I got a call from the Rescue Coordination Centre assigning us our next task. We were to return to the crash scene to shut off the plane's ELT and recheck for survivors. We also needed to pick up police officers from Blenheim, and winch them to the site, so they could guard it overnight.

After dropping the aircraft pilot and paramedics at the hospital, the helicopter returned to base for refuelling. Steve and John had a chance to catch their breath for a few minutes and get ready for the next task.

I volunteered to be the one to return to the crash site. John warned me that Dean had described the scene as a disturbing, mangled mess so he made sure that I was mentally up for the task. I assured him I was and after a briefing we headed back to Mount Robertson.

BY THEN, I HAD been involved in the emergency services for more than 20 years, and over that time I had seen hundreds of dead people. Accidents, medical emergencies, young and old. Death is just part of the job. So, I didn't think it would be any different to any other job I had been on before. I knew that there were five dead people waiting for me on the hilltop and I thought I was prepared.

Crossing Cook Strait, we were relaxed and speculated about the crash and what might have happened. The pressure was off on this mission — there was no one to save, so the job was a straightforward winch down to the bush.

The Sounds Air plane after it crashed into Mount Robertson.

By the time we arrived at Mount Robertson a no-fly zone had been set up for five miles around the mountain, meaning no other aircraft were permitted to fly into the area. This is a standard safety measure which also prevents the media from flying overhead to get their video and still shots.

When we arrived above the winch spot I was sent on my way. As I dropped beneath the canopy, the wreckage came into view and the distinctive smell of aviation fuel hit my nostrils. I was astonished anyone had walked away from that pile of mangled metal.

After I was safely on the ground John gave me a quick radio call, ensuring I was all good, and reminded me that they were heading for Blenheim to pick up the police officers. They would be back in 30 minutes or so, he said.

I removed my helmet and took it all in. In front of me was a crumpled aeroplane which had come to rest in the middle of a stand of trees. It was not immediately recognisable as an aeroplane, more shredded metal, partially wrapped around trees. There were bits of branches, bits of plane and personal belongings from the luggage, which had been ripped open, spread over many metres. I could see the bodies of the passengers still within the fuselage.

I looked up into the trees, trying to figure out exactly where the plane had crashed through. The broken branches high above me were the only indication — the foliage had completely hidden the plane. Without the ELT, the crash site would have been almost impossible to find.

I headed to the plane, ready for my first task — recheck the five passengers for any signs of life. It had now been several hours since the accident, and all the signs of death were present. Pale, cold, clammy skin and eyes staring, unfocused, into the distance.

Then there was the smell. It is indescribable, and all I can tell you is that it smells like death. Death and Jet-A1 fuel. Despite all the obvious signs, I carefully climbed through the wreckage, checking each of them for a carotid pulse. None was present on any of them.

I moved on to my next task, shutting off the ELT located in the tail of the plane. All aircraft have a sticker which says 'ELT Here' to help rescuers find the device.

Using a rock, I broke the skin of the plane and made a hole large enough to stick my hand through. I felt around and could not locate the ELT so I made the hole bigger so that I could get a good look inside.

I discovered that the ELT was mounted on a strut, just above the hole I had made. I changed the frequency on my portable radio to 121.5, the frequency it transmits on. The sound emitted is a piercing siren which continually emits a high and low wail.

I reached up and switched it off; the wailing siren from my radio stopped. Everything became eerily quiet. It was at this moment that I began to get a touch freaked out.

I had this strange feeling that I was being watched. Even though I knew it was not possible, I became convinced that one of the passenger's eyes were following me as I moved around the crash site. I was so convinced that I found myself rechecking their pulse a few times. But there was no pulse. It was just my overactive imagination.

Something I had never realised before was that there is always some other living person around when I am at work. At a car accident scene, no matter how many people might be injured or dead, there are other people and lots of noise. Fire truck engines, emergency service radios, people talking, people doing things. Lots of activity and lots of noise.

But here I was, on top of an isolated hillside, alone beside a wrecked aeroplane with five dead people — and total silence. It made me wonder how people work alone in mortuaries. It also made me appreciate how having other people around helps to deal with the grimness of accident scenes.

I had completed my two tasks and there was nothing else to do but wait for the helicopter to return, so I wandered around, taking in the scene and wondering what had caused the plane to crash.

About 15 minutes later, I heard the distinctive sound of our BK117 in the distance. I then got a call on the radio letting me know that they would winch down the two police officers followed by their gear. Once they were settled in, I would be winched back up.

John's voice had never sounded better to me — I was glad to be back in touch with the living.

Soon the police were in place and I was back in the helicopter heading home. The guys asked how I was and I honestly told them I was fine. I described what I had seen and done, had a laugh about the size of the hole I made in the aircraft to find the ELT, and described the feeling of being freaked out.

When we got back to base, I caught up with Dean. We had a chat about the scene, how amazing it was that anyone had survived, and about how we were doing. We both said we were okay.

Nonetheless, for many days after that, Dean and I would touch base with each other daily, talking about the weather, how the survivor was doing, or what we had been up to. Our shared experience, even though we experienced it at different times, created a special bond between us, which continues to this day.

A bit of sunshine to come out of that bleak day was that Pete was offered the opportunity to become a part-time crewman on the helicopter. His actions impressed the others so much that they wanted him on the team. He did good!

TWENTY
HELMET AND WINCH CAM

IN THE EARLY 1990s, as our media programme grew, I set up good working relationships with both New Zealand TV broadcasters, TVNZ and TV3. At first, all we could do to support the relationship was make ourselves available for interviews following a rescue.

In the mid-1990s, TVNZ gave us a handheld video camera that they asked us to use, if possible, during a rescue. But we soon found that we didn't have the time or focus to try and film what we were doing while operating the winch.

Although helmet cam is common now, back in the mid-1990s it was unknown. I spoke to TVNZ and said what I needed was a way to attach the video camera to my helmet so that wherever I looked, the camera caught some footage. They said they had nothing to offer and we should just try harder.

I was having a beer with Ben, one of the TV3 cameramen, and told him that I thought a camera attached to my helmet would work well. He thought it was possible and the two of us went to see Gordon McBride, the TV3 Wellington bureau chief, who thought it was a great idea.

Ben and Gordon concocted a scheme that saw them 'borrow' a small stump camera from the sports department. This was a camera which sat in the wicket and caught shots of a cricket game. Its small size was ideal for what we needed.

Ben combined the camera with a little video recorder and a battery supply to create a small stand-alone system which sat in a bum bag.

He then made a bracket which attached the camera to my helmet, and there we had it, our first helmet cam. Unsurprisingly, the camera never made its way back to the sports department!

Helmet cam revolutionised the way we filmed rescues from the helicopter. Instead of having to try and hold a video camera, now wherever we looked the camera looked.

One of the first things we noticed was how much our heads moved during a winch operation. Our heads, and therefore the camera, would be continually moving all over the place as we watched the winch hook and the clearances around the helicopter. We also found that the camera angle was not always where we needed it to capture the ideal shot.

As time went on we learned that we only needed to keep our head still for a short amount of time, five to ten seconds, to get enough video footage for a news story. To help with the camera angle, we simply drew a small box on our helmet visor. With the visor down, we knew that whatever we were seeing through the box was also what would be filmed.

We began to capture some astonishing footage showing exactly what we were seeing during a rescue. As long as we remembered to attach the camera to our helmet, turn the video recorder on, and get the okay from the person we rescued, we were sure to get a short story on the 6pm news every time we did a winch rescue. Our increased exposure in the media was helping the fundraising team, too; they now had a new tool to show donors exactly how their donation was spent.

The helmet camera worked so well that TV3 consulted with our helicopter engineers to design a video recording system which attached to the rescue hoist. The camera on the hoist was wide-angle and captured most of the action going on below the helicopter. The engineers also wired the camera to our audio system which meant the conversation between the pilot and crew was recorded alongside the video.

Now we could capture footage from two angles, and this gave people a unique bird's-eye view of what happened during a rescue.

The video footage was also an excellent training tool because the pilot and crewperson could review it and see things from each other's perspective. I found it invaluable for me because I heard when I strayed from standard winch patter. The footage also quickly laid rest a lot of 'you said this' or 'you did that' arguments between the pilot and crewperson. It was also an excellent means to show new winch operators or paramedics what to expect.

Thanks to smaller and better quality video cameras, helmet cam is now obsolete, replaced by cell phones and GoPro cameras. TV today is full of reality programmes which capture footage from every conceivable angle. Our own *Life Flight* reality TV series used a slew of cameras set up in the aircraft and on the crew and paramedics.

It is fun to reflect that we were one of the first helicopters in the world to have this kind of technology, all thanks to an idea concocted over a beer and backed up by some creative Kiwi ingenuity. Gordon and I often joked that we had missed our chance by not patenting the idea!

TWENTY-ONE
TERMINATOR

IN THE BUSINESS world, if you miss a deadline or forget to back up your PC, the consequences can be dire, but they are unlikely to be fatal. On a rescue helicopter, however, when things go wrong the consequences can be deadly for us and the people we are trying to save. Even with all our training, procedures and well-maintained equipment, things can go wrong — sometimes horribly wrong.

It is impossible to plan for every imaginable scenario so continual training is vital, as is ensuring the helicopter and equipment are fit for purpose. Thinking on our feet is crucial as when it is all going pear-shaped, the ability to remain calm and, literally, on the fly come up with Plan B, Plan C or as many as are needed to achieve a rescue is critical.

Saturday, 27 December 1997, was one of the days when this ability was needed. It started as a busy but ordinary day for the team with a flurry of rescues that kicked off with a pre-dawn call-out at 4am.

Steve was the pilot, John the crewman and Dean the paramedic on the helicopter heading out to an accidental shooting at Pelorus Sound in Marlborough. It was a straightforward rescue and the victim was winched from the bush and in Wellington Hospital by 6am.

Steve continued as the day-shift helicopter pilot and I took over as the crewman. Around 11.30am, Kevin Smythe joined us as the paramedic for a mission to Tora, a tiny settlement in the Wairarapa, where an elderly man had fallen on rocks.

It was another uneventful job and he was in Wellington Hospital by 1pm. By 2pm the helicopter was clean and restocked, the paperwork done, and we were contemplating what to have for lunch.

But the day wasn't over yet and at 4pm, with paramedic Iain Mackay on board, we headed to the scene of a major car accident at Pukerua Bay on the Kapiti Coast.

It was a hell of a smash, with a lot of injured people and several still trapped in their vehicles. While the Fire Service worked to cut people free, we ferried the most seriously injured person and another less injured to Wellington Hospital before heading back to collect two more.

As we headed back, Steve and I agreed we were ready to put our feet up and have an afternoon nap after we returned to base when John, who was manning the emergency line, called us.

He had been advised by the Rescue Coordination Centre that several yachts taking part in the Wellington to Akaroa race were in trouble. The race takes the boats across Cook Strait and down the east coast of the South Island, a demanding stretch of water at the best of times, but the wind and weather had proven to be stronger than forecast. The race had almost been called off due to concern over the conditions, but the yachts headed out of Wellington Harbour with a brisk nor-wester at their back, which had swiftly built to gale-force winds.

Strangely, instead of a specific task, we were asked by RCC to take on as much fuel as possible and head south. We would be given more details as soon as possible.

We dropped our last two patients from the car accident at Wellington Hospital and returned to base to set up for a marine search and rescue. Steve was continuing as the pilot, John came aboard as the winch operator, and I was to be an observer and/or swimmer, ready to go down on the winch, if required.

Although no marine rescue is ever the same, by this time I had done many of them so I had few concerns as we headed out of Wellington into Cook Strait at about 5.30pm. Being summer, we had about three and a half hours of daylight left although the helicopter only held about two and a half hours' worth of fuel.

The wind had hit a constant 60 knots, which is considered a Category 1 storm with winds about 119–153 kilometres per hour (74–95 mph). It was right on our tail, so it was going to be a fast trip south!

A few minutes into the flight, RCC rang with the details of our rescue. A crewman had been injured and needed to be winched from a yacht lying roughly 100 kilometres southeast of Wellington and about 28 kilometres offshore from the South Island.

The yacht, named *Terminator*, was a 40-foot Elliot with nine crew members. It had lost its rudder when the yacht broached earlier in the day, meaning it was unable to steer. The crew had managed to rig up their emergency steering equipment, a required backup for an offshore yacht. Once it was in place, they had started their motor, but the yacht just went around in circles because the emergency rig was unable to provide enough steerage as they were being battered by 110–130 kilometre per hour winds and 10-metre-high waves.

Realising they needed help, the skipper requested a tow to shelter in the Marlborough Sounds. Unfortunately, no aid was readily available. Conditions were too rough for the Cook Strait fast ferry and the bigger ferries couldn't launch a rescue craft in the high seas so were also unable to assist.

A fishing boat was sent from Kaikoura but had to turn back due to the rough conditions.

The police launch *Lady Liz III* was already in Cook Strait helping other struggling boats and didn't have enough fuel on board to get to the scene and then safely return to shore. They would have to return to port and refuel before they could head out to assist *Terminator*.

The crew was streaming sea anchors to slow down the boat's drift as it floundered in the heavy seas. But with no steerage, the yacht was rolling dangerously in the big waves.

Several hours after requesting assistance, a massive wave knocked the yacht upside down. It had righted itself, but one of the crew had been thrown around inside the cabin and knocked unconscious.

The crew contacted RCC again and requested assistance for the injured crewman. Since we were already in the air, RCC directed us their way. A few minutes later RCC called and told us to abandon the mission and return to base. After a few choice words and grumbles, Steve turned the helicopter into the gale-force wind and we headed home.

For some reason I've never understood, the yacht's crew was told by RCC that if they wanted a helicopter evacuation for their injured crewman they would need to declare a Mayday right then. The crew were confused by the need to declare the Mayday. Surely an injured person was reason enough to send the helicopter? The crew reasoned that the yacht was not sinking and, by this time, the injured man had regained consciousness and said he could stay on board, so the crew didn't feel declaring a Mayday was warranted. This was when we were stood down.

The yacht continued to be pounded by the seas and the crew debated whether a Mayday was necessary. There were two separate wave patterns and the yacht was being knocked over 70 or 80 degrees by waves every few minutes, so some believed another rollover was only a matter of time. They knew that if the yacht did roll again they could lose their mast, and with the approaching darkness and increasing distance from land, this was becoming a recipe for disaster. The debate continued until the injured crewman became unusually quiet. Now there was agreement that a Mayday had to be declared.

By this time, we were back at base and refuelled, ready to go. A quick weather check and briefing, and we were off.

Given the challenging and deteriorating conditions, we discussed the best options for getting the crew safely off the yacht. Sea rescues in a severe storm like this one are dangerous. We would have to contend with constant high winds, unpredictable gusts and heavy seas. The yacht was lying broadside to the waves so its mast and steel rigging would be flaying about, a threat to the helicopter if it got too close. Add to that the difficulty of the people moving about on a heaving deck. It also meant that I could be injured if I was winched aboard. We agreed that I would only be winched onto the yacht if there were no other option.

As we neared the yacht, John contacted the crew by marine radio. They advised us that the injured man had been briefly knocked out, but fortunately was now conscious and alert. John briefed them on how the rescue would proceed. The winch hook with two rescue harnesses attached would be lowered to the yacht. The harnesses are like a large horse collar that are connected to the winch hook which is then locked. Once on deck, the sailors would have to put them on over their head and under their arms.

WHEN BOTH WERE in their harnesses, they were to give a thumbs-up to indicate they were ready to be lifted. We would pull them off the yacht and winch them to the helicopter. John told them to expect to be dragged through the sea after they came off the deck. He advised them that we would pick up four people on this flight, return to base for fuel, and then come back and rescue the final five. They read back the instructions calmly and clearly, increasing our confidence in them.

Maybe this would go all right, I thought.

As we approached the reported GPS location of the yacht, all we could see was churning white water and spray. We were only 100 metres above the waves and being knocked about by the strong wind. It was clear this was going to be a tricky rescue, but we couldn't start until we found the yacht!

We called them on the radio and they reconfirmed their position, but we couldn't see them and they couldn't see us, so we asked them to activate their emergency locator beacon.

Seconds later, we located them a short distance away. The yacht was rising and falling in the waves, and even after we spotted them, it was proving hard to keep the white yacht in sight amid the foaming white water and spray that engulfed them.

Steve took us down, and once we were in position, John began lowering the rescue harnesses. However, it quickly became apparent that this wasn't going to work. The lightweight harnesses were being blown about in the high winds and were not going to make it to the yacht. We quickly discounted the idea of using our hi-line, a rope with a weighted sand bag at the end, because the high seas meant the helicopter couldn't remain within 50 metres of the yacht and that was the length of the rope.

Normally, we would position the yacht forward of the helicopter and off to the left: at the 11 o'clock position on a clock. This time, John asked Steve to reposition the helicopter further upwind of the boat so that he could try something new. John re-lowered the harnesses and, sure enough, they drifted towards the yacht — a technique we hadn't used before — but proving the theory that we always needed to think on our feet and come up with as many new plans as required.

After a couple of minutes, and with about 36 metres of winch cable played out, the harnesses reached *Terminator*. Two of the crew put the harnesses over their heads, pulled it up under their arms, and gave a thumbs-up. John got Steve to manoeuvre the helicopter back towards the yacht while he winched the cable in as fast as possible.

When the time was right, John called for Steve to pull up and away from the yacht. As expected, the crewmen were briefly dragged through the water before they began rising towards the helicopter.

However, as they came towards us, John and I both noticed that the two sailors didn't look overly comfortable, and something seemed off with the way the harness sat under their arms.

We got them on board safely and then tried to figure out what was wrong. We assumed we had not given them enough time to correctly position the harness. As John began sending the harnesses back to the yacht, I radioed them and advised them to take a bit more time to ensure that it was secure under their arms.

The two rescued sailors watched out the rear window of the helicopter in dismay at the sea state that they could now see from above and wondered aloud how we could carry out a rescue in such severe conditions.

It took significant coordination, skill and teamwork between John and Steve to get the harnesses back to the yacht. This time, John waited longer to ensure that the sailors had time to get them on correctly. Nonetheless, as they were heading up, John could see something was still wrong. One of the guys, who we later found out was Chris Webb, was clearly uncomfortable and his legs were flailing about. John asked Steve to lower the helicopter as close to the sea as possible.

I was next to John lying on my stomach with my head out the door watching the rescue while the video camera on my helmet recorded the events unfolding.

When the two crewmen were only a metre below the skid, the unthinkable happened. Chris suddenly went limp, his hands dropped from the hook, his arms came over his head and he slipped out of the rescue harness plunging about 10 metres back into the turbulent sea.

Despite the momentary feeling of horror, my overriding concern was to keep him in sight, but he immediately disappeared below the surface among the churning waves.

John stayed calm and told Steve what had happened. Our priority was to get the other sailor on board. While Steve turned the helicopter to try and keep track of where Chris had fallen into the sea, John continued to winch him in. He came on board looking shocked and lethargic. We got him into the cabin and instructed the two sailors already aboard to care for him so we could focus on finding Chris.

Video grab of Chris Webb as he falls out of the harness into the sea.
TV3/Life Flight

Video grab of *Terminator* in big seas.
TV3/Life Flight

Terminator **133**

Dave pulling a rescued diver into the helicopter. This is the type of harness that was being used in the *Terminator* rescue.

Although we didn't show it to the rescued sailors, John, Steve and I were deeply upset at what had happened. This had never occurred before, and clearly it wasn't supposed to happen now — we were there to save people, not make things worse!

I didn't want to say it aloud, but it had been difficult enough spotting a 40-foot yacht in these seas, so how were we going to find a single person? To make matters worse, we had not seen him resurface after he hit the water. This was going to be harder than searching for a needle in a haystack because, unlike the needle, there was no guarantee Chris would be on the surface and even able to be found!

We had moved several hundred metres away from the yacht while bringing the third man on board, so we worked our way back and tried to figure out where we had been in relation to it when Chris fell into the sea. John searched from the left of the helicopter while I searched from the right, directly behind Steve. While we continued searching, I radioed the yacht. The conversation was brief — could they see the crewman who fell from the harness into the sea? 'No.' And could they estimate where we were in relation to them when he fell? 'Around where you are now.'

The Maritime Operations Centre (MOC) in Wellington heard our end of the radio conversation, and as this was the first they knew we were having a problem, a few seconds after I finished talking with the yacht they called asking for an update.

There was nothing anyone else could do to help us; either we were going to find Chris or not. The seas were too rough to send rescue boats, and by the time another helicopter reached us, it would be too late. I responded that we would update them as soon as we could and then lowered the volume on the marine radio. Right now, outside chatter would just be a distraction.

We were frantically trying to find Chris, but the conditions were atrocious and fuel was running low. Steve told us we had ten more minutes of search time left before we'd have to head back to Wellington. I don't think I had ever felt so helpless during a mission — this was not how it was supposed to go!

Then something caught my eye at the top of a wave but disappeared as quickly as I saw it. I had glimpsed a bit of orange — Chris's survival suit — popping up on top of a wave and then disappearing into the trough. 'Target 2 o'clock,' I called into my microphone, pointing my arm towards the target, the standard procedure that helps us keep track of where something is, even if I couldn't see it at that moment.

Neither Steve nor John had spotted what I saw but moved to where I thought I had sighted him. Seconds later, Steve and then John saw him as well. It was Chris and he was conscious and waving at us. He was alive!

But the clock was ticking; fuel was getting low and time was running out to complete the rescue. We decided to lower me into the sea without disconnecting from the winch hook. If I came off the hook in those seas, then it was likely they would lose sight of us, and it would be almost impossible to reconnect us again. We had to do this now or the helicopter would have to return to Wellington leaving Chris floundering in the sea. That was not an option.

I removed my helmet, attached the winch hook to my harness, gave John a thumbs-up and was out the door. I had been in the helicopter looking down at waves this big before, but I had never been sent into seas this big. Then the adrenaline kicked in and there was nothing else in my mind — I was totally focused. I kept my eye on Chris as best I could, but as I was spinning around on the hook, keeping him in sight was hard.

Suddenly I was in the water. Huge waves of cold salt water slammed into my body as I was dragged towards Chris — the immersion suit offered little protection in these conditions. As I came through a breaking wave, there he was, directly in front of me. He reached out for me as I reached for him, and then we were together. Later he told me the first thing I shouted at him over the howl of the gale and helicopter engine was the same thing I always say — 'G'day, I'm Dave.'

I told him I was going to put him back in the same harness he had fallen from, but I was with him, and there was no way I was going to let him fall out again.

The most dangerous part of being in the water and being attached to the winch hook is winch cable management. Since the waves were so big and the helicopter was being blown around so much, John had to leave me with enough spare cable so I wasn't pulled out of the water or away from Chris before I was ready. However, the winch cable is relatively thin and made of steel; if any of the excess cable accidentally got wrapped around a limb or a neck and the cable went taut, it could maim or kill us. I managed to keep the cable clear as we got the harness in place around Chris. Our heads would stay on our bodies for now!

I locked my legs firmly around Chris's legs and gave John a thumbs-up, indicating that we were ready to be pulled up. I carefully managed the cable and as the slack disappeared, we were dragged through the water towards the helicopter. As we emerged from the waves, I started to tell Chris that we'd be okay now but before I could finish my sentence we got hit by a wave. I shut up at this point, keeping a firm hold on him as we spun around, making our way up to the helicopter. Thirty seconds later we were safely at the skid, and then made our way into the cabin.

The look on John's face clearly reflected the relief that we were all feeling at that moment. As I came through the door he had a huge grin, and I wasn't sure if he was going to hug me or kiss me – this was one of the scariest moments of the day! He settled on a manly pat on the shoulder, and once we were off the winch hook, John secured it, shut the door and we were on our way home.

John radioed the yacht and MOC to let them know that we had found Chris, he was okay, and we were heading back to Wellington for fuel. He also requested an ambulance to meet us to check the four sailors we had aboard.

Remarkably, our job was only half done. The three of us were tired and ready for a break, but we had five more people to rescue. But before we did that, we had to figure out what had gone wrong.

Our rescue harness had been specially developed to rescue unconscious people from the water. As the harness tightened, it was supposed to lock around and keep hold of the person, especially if they couldn't hold on themselves.

All four of the crew reported the same problem. They told us that their lifejackets had been driven up into their throats and choked them. When we examined their lifejackets, we found that they were very thick and didn't have a crotch strap — which would pass from the front to the back of the lifejacket between their legs and prevent the jacket from riding up.

We surmised that our rescue harness and the offshore lifejackets were incompatible. As our rescue harness tightened around the lifejacket, it forced it up to their throat. All of them experienced this, but in Chris's case it choked him to the point of unconsciousness. As everything was saturated, when Chris passed out, his arms raised above his head and he slipped out of the harness.

Now that we thought that we understood the problem, we figured out a work-around, but we were not sure the remaining five sailors would like it.

It was a slow trip back home into the headwind. When we got to base, we helped the crew off the helicopter and refuelled. I was soaking wet so jumped into a new immersion suit, and within 20 minutes of landing we were heading back to the yacht.

As soon as we were close enough, John radioed the yacht and ran through what we thought had happened, and why, and how we had solved it.

'You guys need to work together on this. Just before you get into the rescue harness, you need to remove your lifejacket. If you come up to the helicopter without a lifejacket I guarantee you will not fall out of the harness; however, if you get swept off the deck between removing your jacket and getting in the harness, then you will probably not be able to be rescued.'

John released the transmit button and we waited almost 15 seconds for a response, and when it finally came, it made the three of us laugh out loud: 'Pardon?'

We assumed that they had misheard John with his English accent, so I gave the briefing again in my American accent. We could hear the doubt and worry in their voices, but they agreed that they would do this.

We tracked in on the emergency locator signal and part two of the rescue began. Steve and John were well practised at this by now and it went smoothly. The first two people came up without any signs of discomfort. Since there were now three of them left, the next winch up was with only one person — that way the remaining two could assist each other into the harness. On the third winch, the last two came off the yacht.

With darkness approaching, we headed back to Wellington, landing at the base around 9.30pm. By now there was a party happening there. Some friends of the yacht crew had arrived, as had TV news crews and newspaper reporters. After we cleaned the helicopter and refuelled it, ready for the next mission, I had a quick shower and change of clothes and then joined the party.

It was a huge day and by now I was exhausted, but never had I ever been so relieved. Chris was relatively unscathed, apart from a chipped vertebra, presumably from hitting the water hard after falling to the sea below.

There was even a happy ending for the yacht. The emergency beacon was still transmitting so several days later a fishing boat tracked the signal and recovered it for the owner.

In the days and weeks following the *Terminator* rescue, we had a close look at everything that had gone wrong and right on the day.

We sourced a new rescue harness, nicknamed a 'nappy' for ship and yacht rescues. With this model, the person being rescued sits in the harness instead of putting it over their head, so that the rescue harness and a lifejacket can't affect each other.

More importantly, we had nine grateful sailors happy to have survived their ordeal. Chris and I became good friends and he buys me a beer every time we get together. Four years after the rescue, I was invited to Christchurch to help celebrate Chris's fiftieth birthday. Surrounded by many of his friends and relatives, someone made a toast to the helicopter crew who ensured that they all still had a son, father, uncle and friend.

It was a special celebration and again reminded me what a difference our team made, day in and day out.

TWENTY-TWO
RESPONDING TO THE SCENE

ONE OF THE riskiest parts of the job was landing near the scene of an accident or medical emergency. Responding directly to the scene saves valuable time getting our paramedics to the patient, or the patient to hospital. Sometimes a few minutes could be the difference between life and death. However, this usually meant landing in an unfamiliar or unsecured area.

Everything between us and our landing zone (LZ) as well as anything on the ground could be hazardous to our team or people on the ground, and that included overhead wires, fences, trees, branches, lamp posts, loose debris, tents, laundry hanging on a line, outdoor furniture, TV aerials, roofing iron, people and animals!

I had my first experience of how dangerous landing in an unsecured area can be just a few months after I started crewing. We were called to a lifestyle property in Kapiti to transport a person seriously injured in an accident. The ambulance officers at the scene identified a paddock close by that we could land in, and warned us of the nearby overhead wires.

We circled above the paddock before committing to land and everything looked okay. The nearest trees were 40 metres to our left so as I sat in the open door most of my attention was on the tail rotor, and the ambulance crew working beyond it. Suddenly, when we were only 7 or 8 metres above the ground, Spence, my pilot, aborted the landing.

As we hovered above, he pointed out a very thin wire sitting about 5 metres above the ground, directly in our landing path. I struggled to see it until he moved the helicopter, changing the angle of the sun. Now, there it was, clear as day, glinting in the light. The importance of being vigilant when coming in to land was crystal clear and I carried that lesson into the rest of my career. Had Spence not spotted the wire, we certainly would have hit it and I probably would never have had the chance to learn the lesson.

When a helicopter lands, it creates tremendous downwash which is strong enough to blow over fences, lift loose roofing iron or send anything unfastened, including laundry hanging on a line, tents or ambulance blankets flying. The flying debris could hit and injure someone on the ground, or be swept up into our rotor blades, causing us to crash.

There were many ways we mitigated the risk of scene landings.

At the very top of the list was good, clear Crew Resource Management (CRM); everyone on board working together as a team, keeping our eyes looking outside the helicopter as we landed or took off.

We also worked closely with the emergency services and ensured they knew how to identify and secure an LZ. Ideally, we would talk to a police officer or firefighter at the scene before we approached for landing. They were our eyes and ears on the ground and would suggest where they believed was the safest place for us to land.

One day we were getting ready to land near a road traffic crash in the Wairarapa. As we neared the ground, pilot Tim Douglas-Clifford called 'going around' and lifted back into the air. He explained that he felt a bit of a tailwind and it looked like a forecast wind change had arrived early. The firefighters had done everything right, but instead of using the LZ they had chosen we landed in a paddock nearby. The paddock gave Tim more options when it came time for take-off.

But having an identified LZ was not always possible. In some of the more remote parts of our region we would often be sent 'first response', arriving at a scene before anyone else got there.

Over the years, we built up a list of landing zones in communities around the region. Having a known LZ was a significant safety advantage because we got to know where the wires and other overhead hazards were, reducing some of the dangers we faced.

Regardless of whether an LZ had been established or not, we always remained vigilant, approaching every landing slowly with the pilot ready to fly away if any dangers were spotted.

Helicopter landing on State Highway 1 in Kapiti. *Dave Chapman*

TWENTY-THREE
SEPTEMBER 11, 2001 — NZ TIME

SEPTEMBER 11, 2001 is a date forever burned into the memories of people who were old enough to understand what happened that day. It's a day that changed the world; it's also a day that profoundly affected my best friend's family for a very personal reason.

Wellington is the first capital city in the world to see a new day. It is 12 hours ahead of Greenwich Mean Time (GMT), and in September we were 18 hours ahead of New York time. By the time the first plane crashed into the World Trade Center's North Tower on the morning of September 11, I had already had a crazy day in New Zealand.

I was off flying duty but was at work catching up on paperwork when around lunchtime the helicopter was called out for an urgent transfer of a patient. About 35 minutes later, ambulance control called asking if our backup helicopter was available for a 'triple A' — a severe medical condition — in Paraparaumu, a 15-minute flight north of Wellington.

The aorta is about the thickness of a garden hose and is the major artery supplying blood to the lower part of the body, running through the chest and abdomen. An abdominal aortic aneurysm, or 'triple A' (AAA) occurs when the aorta becomes very large or balloons out. If the aneurysm grows large enough it can leak, causing internal bleeding and the person will bleed to death if the bleeding is not stopped. Due to the location of the clot within the abdomen the only thing that can save a person with a leaking AAA is urgent surgery to stop the bleeding and repair the artery.

Time is always of the essence when someone has a leaking AAA. Even if the patient gets to surgery before they bleed to death, the chance of them dying is greater than 50 per cent.

As I was pulling the helicopter out of the hangar, my cell phone rang and I saw it was my best mate Doug. I answered and before he could say a word, I told him I couldn't talk because I was getting ready to go flying.

As I was about to hang up on him, he yelled out: 'WAIT, IT'S URGENT!' Doug is one of the people who understands, better than most, what it means when I say I'm going flying, so for him to yell out like that stopped me in my tracks.

Doug had just had a call from his mum who was in a panic. His dad had not returned from running errands and she had heard that he had collapsed and an ambulance had been called. She couldn't find out anything and he hoped I might be able to get more information.

I got a horrible feeling as soon as Doug told me his dad was in trouble, and worried that it was his father John that we were going to pick up. I didn't share this with him, because I didn't want to cause him more anxiety in case I was wrong, but I am sure he had the same feeling.

This was quickly changing from being a straightforward mission to a very personal one. Doug and his family were close to me. I had been to many of their family and social gatherings, and I had stayed at his parents' house several times. In New Zealand, the family connection didn't get any stronger for me.

I promised Doug I'd see what I could find out. I called ambulance comms and it didn't look promising; the only serious job going on in Kapiti was the one we were responding to.

As soon as David Boyadjian, the paramedic, arrived we headed to Coastlands Mall. I texted Doug and told him where we were headed and that I would let him know as soon as we landed if the patient was his dad.

Dave, aged eight, visiting the fire boat at his local FDNY station.

Dave operating the winch during training in the bush.

Harry Stevenson and Dave winch the stretcher to emergency crews waiting on the Rimutaka Hill. *Dominion Post*

Two WFA paramedics during winch training.

Tim Douglas-Clifford and Ruth Zeinert winch paramedic Iain MacKay to an injured person on rocks.

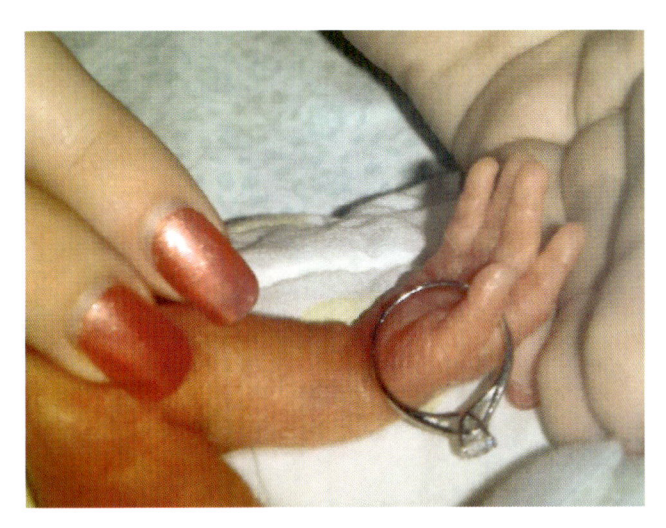

Deanna Sigmund's ring shows just how tiny her premature baby, Cameron, was.
Ben and Deanna Sigmund

The helicopter landing at the scene of an accident. *Derek Quinn/111 Emergency*

The helicopter landing at the scene of a road traffic crash near Grovetown. *Marlborough Express*

Mike Hall and Jon Leach winch paramedic Keith Frewen
and a patient to the helicopter. *Marlborough Express*

Emergency services above Colin Tandy's truck, which rolled down a bank.

Harry Stevenson and Dave winch paramedic Aaron Hartle and patient Colin Tandy to the helicopter. *Marlborough Express*

Katie-Jane Bowen, motorcycle crash survivor, with Mike Hall, Tor Riley and Dave.

Colin Larsen outside the helicopter as it comes in for a night-time landing. *SNPA/Ross Setford*

Logging and helicopter crew carry a patient to the helicopter after an accident.

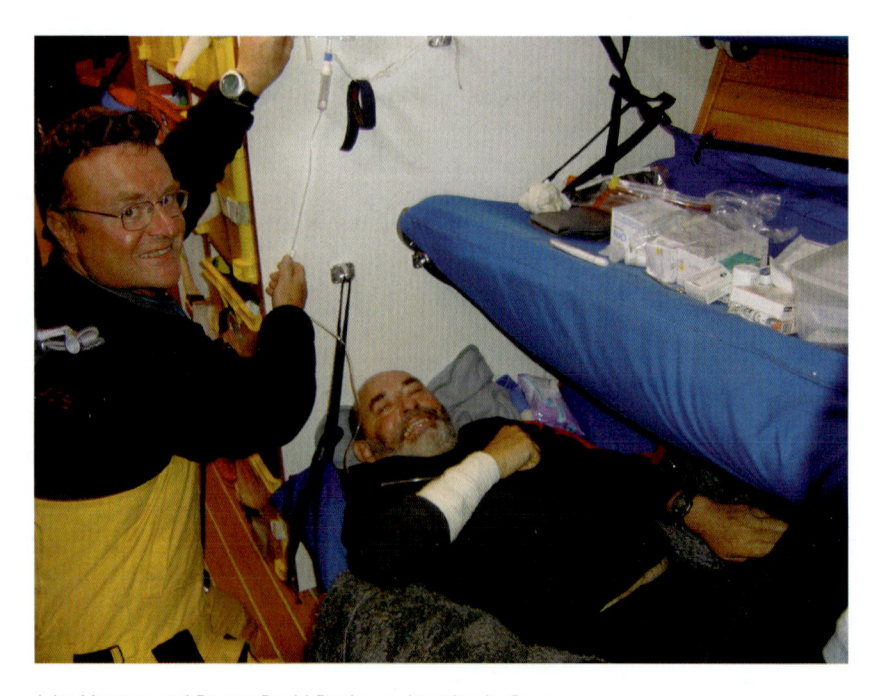

John Masters and Doctor David Roche on *Imagine It. Done.*

The helicopter over *Imagine It. Done.* as seen from the spotter plane.

John Masters and Iain being winched up to the helicopter from *Imagine It. Done.*

What we see through night vision goggles.

Dave with his mum, dad, sisters and their husbands.

Photo of the whole USA family together in 2013.

The view as we arrived in Christchurch two hours
after the earthquake struck on 22 February 2011.

Aerial view of St John's triage set-up at Latimer Square
following the 22 February earthquake.

Firefighters cut off the damaged skids and built a bed of tyres so our helicopter could land safely after its crash.

Life Flight's BK117 cockpit. The pilot sits in the right-hand seat, crew person on the left.

Mike Hall and Dave winch a volunteer firefighter during a training drill. *Marlborough Express*

The helicopter landed on a bulk carrier ship in the Tasman Sea to uplift an injured sailor.

The helicopter on the ground during the rescue of an injured tramper on Mt Tapuae-o-Uenuku in Marlborough.

The helicopter searching for a missing swimmer in Wellington Harbour. *SNPA/Ross Setford*

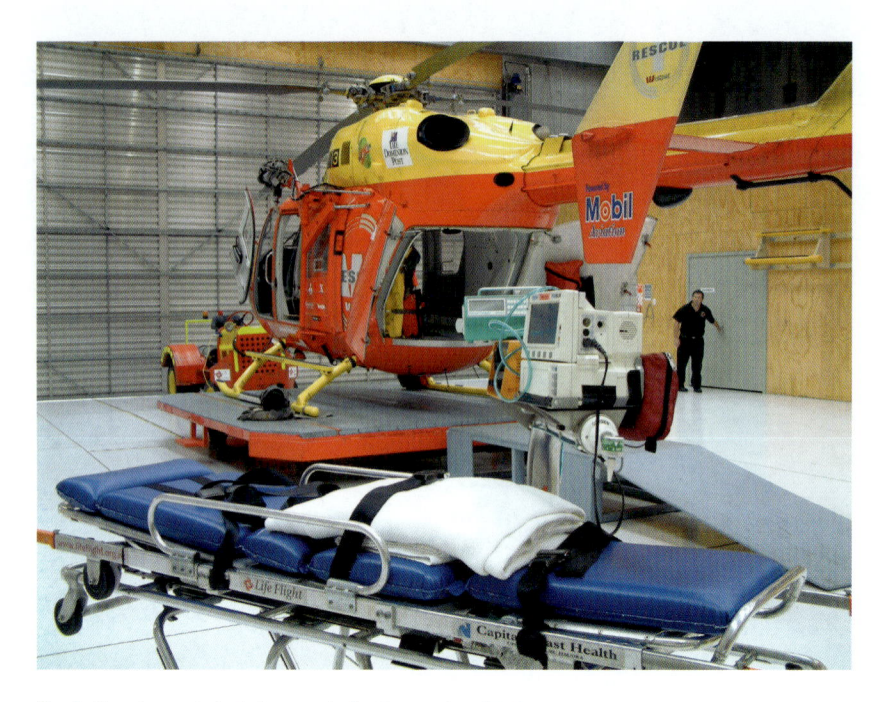

The helicopter and stretcher ready for the next call-out.

Logan Taylor, Roger Hortop, Tim Douglas-Clifford, John Goldswain, Grant Withers, Dean Voelkerling, Julian Burn and Dave at Dave's farewell as the Operations Manager, May 2014.

It's a horrible feeling to know that a patient is someone you are close to, but our most important task on any mission is safety — and that means staying focused. We were landing in the parking lot of the biggest shopping mall on the Kapiti Coast. In addition to the usual hazards we face with an off-airfield landing, a helicopter is a magnet that attracts people like bees to honey.

A helicopter does not care who our patients are or, for that matter, whether we are on a scenic ride or an ambulance mission and a helicopter does not care what we are doing. It has spinning rotors which will kill someone if they walk into them. The Squirrel helicopter is potentially more dangerous than the BK117 because the tail rotor is much closer to the ground, making it easier for someone to walk into it. Therefore, it was essential that we had a secure landing zone and that I focused on my job when we landed.

As we circled above the mall, we saw the fire brigade had a good landing zone set up for us and the area well secured. As soon as the skids touched the ground, I leapt out and went around back to guard the tail rotor — this is a critical task until the rotors stopped spinning. But, I didn't want to be guarding the tail rotor — I wanted to find out if our patient was John. Yet my training was drilled into me — safety first. I waited for the rotors to stop spinning.

As soon as I could, I dashed across to where the paramedics were working on the patient. My fears were confirmed, John was lying on the ground, pale, sweaty, with a swollen tummy and in intense pain.

Any ambulance officer in the world will tell you that one of the best bits of medical care we carry with us is compassion. A reassuring smile, holding a hand or touching a shoulder are the best things we bring to some patients. That is not the case with a leaking AAA. The patient needs plenty of IV fluids to replace the blood draining into their body cavity. They need oxygen. And they need surgery. However, compassion is still good. I made my way to his side, his eyes were closed.

'John, it's Dave,' I said as I took his hand and gave it a squeeze.

His eyes opened and he gave me a bit of a smile. 'Hi Dave, funny running into you,' he said. I was relieved that he was semi-conscious, seemed alert, recognised me and still had his sense of humour. All positive signs.

While the paramedics got John ready for transport, I rang Doug to tell him that it was his dad we were picking up and that we'd be in the Wellington Hospital Emergency Department in about 25 minutes. I told him we were doing everything we could and he should meet us at the hospital. Doug's voice trembled a bit, a sign of how upset he was by the news. He asked a few questions, looking for reassurance that his dad would be okay, but, as much as I wanted to, I couldn't promise him anything. All we could do was get John to hospital as quickly as possible.

When the clock is ticking, especially against a member of your family, even the smallest delays are upsetting. The Squirrel is a much smaller helicopter than the BK so everything takes more people and more time. Instead of rolling a stretcher straight into the back and having it automatically lock in, four people must carry it to the helicopter, manhandle it in place and then seatbelt it down. The Squirrel is not an ideal working space if the patient needs a lot of care in the air.

David sat on the small seat above John's head, and I sat on the bench seat alongside. Once we were airborne, I held onto John's hand while David monitored his condition. John's abdomen cavity was rigid and continued to swell from the blood leaking into it. There was nothing else we could do. I knew fast transport to the hospital was giving him his best chance, but clinically it wasn't looking great.

Once on the hospital roof, again unloading the Squirrel was more complicated and took longer than the BK. I knew time was John's enemy and every lost second was agonising.

As expected, a medical team was waiting when we arrived in the Emergency Department (ED). What was not normal was seeing Doug waiting there, too, looking more upset than I had ever seen him, but he managed to hide all his emotion as he said hello to his dad.

After just a few minutes in ED, John was sent upstairs where a surgical team was preparing for immediate surgery. At times like this, my red helicopter suit and a set of hospital access cards were useful as it meant Doug and I could get into the lift and escort his dad up to theatre level.

There is a staging area between the lifts and the sterile operating theatres where many things, including paperwork, are completed. As we arrived, I saw a familiar scene unfold, but for Doug it must have looked like chaos — five or six doctors and nurses in surgical gowns, talking to each other, talking to John while they poked and prodded him, filling out forms — all requirements before surgery.

At one point a doctor came over to speak to John: 'Do you understand what is happening?' he asked. 'I need you to sign this form. Do you understand that there is a 50 per cent chance you might not make it?'

'Don't you dare talk percentages to him,' said Doug, strongly and clearly.

That made us all pause and take notice.

I had heard doctors speak to patients and their families many times before. Some doctors have a better way of presenting news than others, but it is their job to be honest and not to build up false hope. Maybe the news was not delivered in the way someone would like to hear it, but they needed to know. You are very sick. You might die. The surgery might help. It might not.

But Doug was having none of it. 'Dad, you are going to be okay, I will see you after the operation,' he said, squeezing John's hand.

The doctor wasn't at all fazed and didn't take much notice of Doug. He got his form signed and moved on to his next task.

Doug and I share a belief that the mind is a powerful thing. If you think you can, you can; if you think you can't, you can't. Either way, you're right. Doug didn't want any negativity in his dad's head. I'm sure he also didn't want to acknowledge that his dad was gravely ill.

Doug's reaction to the doctor's words really struck me, and although I always thought of myself as a compassionate person, from that day on I have tried to choose my words as if I were delivering them to my best friend — or his dad. It helped make me a more compassionate person, never a bad thing.

A few minutes later John was rolled into the operating theatre. We had no idea if this was the last time we'd see him alive. AAA surgery can take many hours. It is a very invasive and complicated operation, with a high mortality rate. It was going to be a long and stressful wait for news.

DOUG'S PARTNER TRISH and their 10-month-old daughter had headed to Paraparaumu to pick up Doug's mum as soon as they knew John was being flown to hospital. Trish had watched the helicopter fly past her as she drove north and she wondered if her baby would ever get to know her granddad. Once they arrived at the hospital, and I knew his other siblings were on their way, I knew Doug had support for the moment, so I headed off.

I went back to work for a while, on the pretence of doing my paperwork, but more just to give myself a little space. Normally an ambulance job is not stressful for me, but right then I was stressed and upset and I wanted to prepare myself for the worst.

An hour or so later, once I was feeling a bit better, and ready for whatever was to come, I returned to the hospital to wait with the family. I had watched the same scene hundreds of times, but now I was part of the group supporting each other, sharing stories and praying that John survived.

Finally, hours later, word came that John was out of surgery and was doing okay. The hugs and tears of happiness flowed. We had all prepared for the worst, but hoped for the best, in our own way.

John was eventually brought to ICU and the family was allowed in to see him. He had a breathing tube down his throat, tubes and wires running around his body, and was swollen and pale. I spend a lot of time around patients looking like this and it never bothers me, but that day it did. If I was troubled by it, I could only imagine how it was making the others feel.

A while later I went home to grab a shower and have something to eat. It had been an exhausting day and I was drained.

Around 10pm, I got a huge surprise when Doug called to say that his dad had already had the breathing tube removed and he was groggy, but awake. This was the best news any of us could have hoped for. Things were looking very positive.

I went to sleep around 11pm feeling happy that I was part of a team that not only did great things, but today we had helped save someone very close to me. It was a great feeling.

As I went to sleep that night, September 11, 2001 was one of my best days ever.

TWENTY-FOUR
SEPTEMBER 11, 2001 — US TIME

AT 12.45AM MY cell phone rang.

A phone call after midnight is not unusual for me but when I saw it was Doug, after the day we had just been through, I knew it had to be bad news. I just had no idea how bad.

When I answered, Doug began babbling something about a plane hitting a building in New York. Utter drivel.

'Doug, are you okay? Is your dad okay?' I asked, convinced that Doug had lost the plot. He was now talking about the World Trade Center. Finally, he said: 'Dave, Dad is fine — get up and turn on CNN.'

Confused, I got out of bed and turned on the TV. Although I knew the TV was tuned to CNN it seemed like I was watching a movie channel, because I was seeing one of the Twin Towers on fire. Underneath was a red banner screaming 'Plane Strikes World Trade Center'.

Minutes later, along with millions of people around the world, I watched in horror as the second plane flew into the other tower.

I started switching between news channels. There was no way that both CNN and BBC could be showing live video feeds of something that was made up. This was real. In New York City. My mind immediately went to my family who lived there.

My dad was still working part-time in Manhattan, not downtown, but in midtown, not too far from the Empire State Building. When the second plane hit, it seemed that New York was under attack and I worried that the Empire State Building could be their next target.

My mum was probably at their home, in the Pocono Mountains, about two hours outside of Manhattan, but sometimes she went into the city with Dad for a couple of days. My sisters both live and work on Long Island so were probably okay, but their husbands occasionally worked in Manhattan. My mind began to race — what about my nieces, nephews, cousins and all the others I knew in New York?

I tried to call my dad but the international phone lines were totally blocked and no calls were getting through to the USA. Every time I tried, all I got was a message saying the lines were overloaded and to try again later. I desperately wanted to find out if everyone was okay, but I couldn't get through.

It was still the days of dial-up internet, so it took me a while before I thought to email him. I sent a simple message: 'Are you okay?' A huge weight lifted off my shoulders when a short time later I received a reply: 'Yes, I'm fine.'

Another quick email exchange confirmed that Mum was in the Poconos and that he had spoken to both my sisters and they were all okay. Over the next few hours, I was glued to my TV watching the events of the day unfold.

I found the most upsetting part of the coverage was watching people jump from the upper floors of the towers. I tried to but could not imagine what was going on inside of the building that so many people decided jumping to a certain death was a better option than staying. Eventually, thankfully, the TV networks stopped showing the jumpers.

Then came the horrifying live shots of the Towers collapsing to the ground, one after the other. My first thought was for all the people, particularly the emergency service workers, who would have been in the buildings as they collapsed.

I knew my immediate family was okay, but I had no idea if anyone else I knew in New York had been in the buildings. I felt I was watching a big part of my childhood collapse — watching the World

Trade Center towers grow into the sky was one of my childhood memories. I sat looking at my TV in shock — it was all so surreal.

Although I had heard from Dad by email, I continuously tried to call Mum because I wanted to hear her voice. Finally, around 3am, my call got through. When she heard my voice, she burst into tears. 'You would have been in one of those buildings,' she sobbed. For the first time ever, I think she was happy that I was living in New Zealand and relieved that I was not a New York City firefighter!

Soon after I hung up, my cell phone rang again. It was Mark Stevens, a good friend and a reporter for the *Evening Post*, Wellington's afternoon newspaper. He started the call with an apology for calling me at 3.30am, and after making sure I knew what was going on, and ensuring my family was okay, he asked if there was anyone I knew in New York City that might speak to him. I gave him Dad's phone number and wished him luck in getting through on the phone. I then emailed Dad to say that Mark might be ringing, and it was up to him whether he talked to him.

A little while later Mark rang to say that he had spoken to Dad, and was going to include the interview in the *Post*'s coverage of the event. In retrospect, one of the few light moments of the tragedy was the ensuing article in the *Post* which included some quotes from 'Bill Greenberg, father of Westpac Rescue Helicopter Crew Chief Dave Greenberg'. This was around the time when our media programme was strongest, but the communications manager at Westpac was still surprised that I managed to get the Westpac name into front page coverage of the single biggest world event in our lifetime.

Later, I managed to speak to my dad and my sisters. It was a horrible, emotionally draining night, after what had already been an emotionally draining day.

Around 6am, I realised there was little chance of getting any sleep, so I headed to the gym. While there, I got a phone call from Polly at 91ZM, asking if she could interview me about what was going on in the USA. During the interview, I finally became emotional about it. While talking with her, it finally hit home what I had been watching overnight. The world was never going to be the same and my hometown was forever scarred, but the people there were coping as only New Yorkers could. Even though I didn't usually think about it, that morning I was proud to be a New Yorker.

I finished up at the gym and a few hours later, after stopping in at work, I went back to the hospital and sat with Doug and the family for a good part of the day.

While I was with them, I realised that they were not talking about the events in the USA very much. Their concern was much closer to home. This gave me another good life lesson — a personal tragedy that affects only one family and their network is more important than anything else going on in the world at the time. For Doug's family, John recovering in ICU was the most important thing in the world. Family first.

John was discharged a week or so later, and went on to live a relatively healthy life for another ten or so years. As close as I had been with the family before, I was even closer now. When Doug's parents arrived at his fortieth birthday party I was overcome with emotion, happy to see his dad there to celebrate another family milestone.

Doug's family went on to support Life Flight in many ways. Tara, who had been just 10 months old when it happened, went on to feature in a TV commercial telling people how her granddad got sick and Life Flight helped save him. The cute little girl who was happy we saved her granddad helped raise a lot of money for Life Flight.

The impact my job has on families became crystal clear a couple of years later when Trish introduced me to a group as 'the reason my children still have a grandfather'. The words sent chills through my body.

The events of September 11, 2001, both in New Zealand and the United States, remind me how important the work of the emergency services is and how what we do affects many more people than just the patient.

TWENTY-FIVE
ROAD TRAFFIC CRASHES

I AM NOT a big believer in miracles, but I have seen people survive accidents which appeared to be unsurvivable.

In one, a driver who was not wearing his seatbelt sideswiped a tree and was thrown more than 20 metres from his car to the side of the road and survived with only a few cuts and bruises. In another, a moment of inattention was all it took for a teenager to have his bicycle clipped by a passing train. Instead of ending up under the train, he bounced off it and although he suffered serious injuries, he survived.

Sometimes, when the universe rolls the dice, it comes up double sixes. But after attending hundreds of crashes, I can tell you that the messages on those annoying road safety TV and radio commercials are true. Seatbelts do save lives; the faster you go, the bigger the mess; and friends don't let friends drive drunk.

Over the past 25 years, I have seen a notable reduction in the serious crashes we attended due in part to improved technology in vehicles and safer roads. A driver who might have been killed when their chest smashed into their steering wheel in 1991 will probably be saved by an air bag today. Instead of relying on two yellow lines painted on the road, wire barriers now separate vehicles driving in opposite directions.

It all helps, but the common denominator is still the driver behind the wheel. When I see a car with a baby sitting on their parent's lap or a young child standing up in the back seat, I want to stop them and tell them about the crash I went to where two young kids were killed because they weren't in car seats. Or, even better, about the 3-month-old who survived because they were strapped in a car seat despite the car rolling several times.

But I never stop them because I am sure that the unsolicited advice would be unwelcome. Yet, the next time I was at a car accident where an unrestrained child was injured or killed, I would wonder if someone, anyone, taking a moment to tell these parents the reality of life and death, would have made a difference.

One of the great dichotomies in my life has been the feeling of dread when I fear I might know the people involved in an accident, and the feeling of relief when I don't.

One morning, we attended an accident where a truck had rolled on a bend, crushing a car travelling in the opposite direction. Minutes before we were called out, I had been texting with a friend, who was travelling as a passenger along this stretch of road. On the

Neither the bike nor its rider were in good shape after being struck by a car.

way to the accident, I sent her a text but didn't get a response and began to worry. As soon as we were safely shut down, I ran over and looked under the truck. I cannot describe how happy I felt when I discovered it wasn't my friends.

I should have thought more about the friends and family of the person who was crushed in that car, but I didn't. Sometimes I wish I wasn't so hardened to this stuff.

One of the challenges I face, as do many in the emergency services, is feeling guilty when someone has come into work and thanked us for saving their life and I can't remember them or the details of their accident. For them, it is the worst day of their life; for me, it is one of hundreds of accidents that blend together in my memory.

When I do recall accidents, it is often because something unusual about it made it stick. One of them was a motorcycle accident I attended in 2011. One moment Katie-Jane was riding her new motorcycle on a road in the Wairarapa and the next thing she knew she woke up, lying in a paddock. She had no idea where she was and her cell phone battery was running low so she couldn't stay on the line long enough for emergency services to find her.

This truck rolled onto a car on State Highway 2, Rimutaka Hill.

A two-and-a-half-hour search, involving friends, police cars, fire trucks, ambulances and our helicopter ensued. Pilot Mike Hall, paramedic Tor Riley and I were searching from the helicopter and were getting more concerned for her safety as it grew dark. We could see the flashing lights of different emergency vehicles all over the district, as all of us frantically searched. Finally, a firefighter spotted a break in a fence and found her and her crashed bike 3 metres down a bank.

The helicopter was only a few minutes away, and we were by her side quickly. She was showing signs of spinal injuries, so we treated her with kid gloves, ensuring we kept her neck and spine straight. On the way to the hospital I held her hand and leaned in closely talking to her, hoping she could hear me over the noise of the helicopter.

At the hospital, they found that she had a broken neck and back, amongst other serious injuries. Despite suffering from ongoing chronic pain, today she manages to keep a smile on her face and offers support to people who have suffered debilitating injuries.

Every time I see Katie-Jane pop up on Facebook I am reminded that circumstances don't define a person, the way they respond to those circumstances does.

Another memorable accident was, ironically, on Tumbledown Road in the Marlborough Sounds. Colin Tandy was driving his truck along the narrow, winding road when he was blinded by sun-strike. He braked hard, but his 16-tonne logging truck swerved off the road, sending it plummeting down an 80-metre bank.

By the time we flew across Cook Strait and arrived above the accident scene, some of Colin's workmates, paramedics and firefighters had scrambled down the hill to help him. Pilot Harry Stevenson and I winched paramedic Aaron Hartle down to the scene and then performed a challenging stretcher winch to bring them back up to the helicopter. Colin suffered severe injuries, including broken hips and a broken back, but has recovered well.

It is hard for me to comprehend that anyone could survive an accident where the forces were so great that a truck could clear a path of fully grown trees as it rolls down a hill. That he survived the accident is a testament to his resilience and the teamwork of all those who came together to save him.

In many ways, we were the luckiest of the emergency service people attending major crashes. We showed up, picked up the most seriously injured person from the crash and then flew them to hospital. Although these people were usually severely injured and, unfortunately, many of them died or ended up with crippling injuries, we didn't have to deal with the aftermath.

Often, as we took off from the scene, I would look down and think about what the firefighters and police officers left behind must deal with. At any crash scene, all the initial effort goes into freeing the people who are still alive. Once the living patients are on their way to the hospital, the grim task begins of separating the people who have died from what remains of their vehicle.

It is an onerous, usually thankless, but extremely important task which is done with great care and sensitivity. Every firefighter I know ensures that the dead are treated with care, respect and dignity.

The helicopter landing at the scene of a road traffic accident near Grovetown. *Marlborough Express*

TWENTY-SIX
INTO THE NIGHT

SOME PEOPLE FEAR the dark. Helicopter pilots who fly under visual flight rules have more reason to fear the dark than most!

On my first flight on the Westpac Rescue Helicopter, looking out into Cook Strait, all I could see was blackness; there was no distinguishable horizon and other than a few fishing boat lights in the distance, it was a black hole. As soon as I turned my head back towards the shoreline, everything seemed better.

Night missions had other problems besides the dark. Fatigue could be a problem, as a flight can be requested before the team has had a chance to go to sleep, or a few hours after going to sleep, which is often worse.

The introduction of night vision goggles (NVG) made night flying much safer, eventually.

The goggles, which are shaped like small binoculars, mount to the top of a helmet and turn a black night into a green-tinged day. They magnify any available light, whether it be the moon, stars or man-made light, thousands of times, allowing the user to see in the dark.

We received our first two sets of NVG in 1997. On the night they arrived, John and I were so excited to try them out that we drove out to Moa Point Road, an unlit, winding, two-way road near the airport, and shut off the headlights! We were both flabbergasted by how clear the road was with the goggles on, but as soon as we took them off it was pitch black.

Helicopter with Nitesun searchlight installed. *Marlborough Express*

They were a revolutionary piece of gear which vastly improved the effectiveness of our night searches. Using them, we could spot someone waving a torch, or a cell phone, or any other type of light, from kilometres away.

In one of the early successes with the NVG, we were searching for a small boat which was overdue on Wellington's west coast, with two people on board who had not returned home by nightfall, which was highly unusual.

We began searching along the coast, with the pilot and one policeman searching the shoreline using the beam of the Nitesun searchlight, while another policeman and I searched out to sea using the NVG.

As we flew along, the two of us using the NVG were sure we saw a small flash of light near Mana Island, about 4 kilometres from where we were. The pilot was doubtful since he could see nothing but darkness. But, he gained some altitude and headed for the island. As we got closer, we definitely saw a torchlight being waved around. The two not using NVGs didn't see it until we were only a few hundred metres away.

Thanks to the NVG, we found the pair hours earlier than we would have using a standard search. We called in a Coastguard boat which towed them back to shore.

The NVG proved to be invaluable search tools but, unfortunately, it would be another 10 years before our pilots could legally use them. It seemed crazy that the crewperson could sit alongside the pilot, describing everything they could see outside, but the pilot wasn't allowed to wear them.

Although military pilots had been using NVG for years, they weren't approved for civilian use in New Zealand for many years. Finally, around 2006, new rules permitted civilian use.

Life Flight spent well over $100,000 on training for the pilots and crew and upgrading our helicopter cockpit to be compatible with NVG.

The goggles have some limitations, and they don't give a pilot the ability to fly in all weather conditions, but they were a major safety improvement because the pilot could now see the terrain around them in very low light conditions.

Today, all the New Zealand rescue helicopters use night vision goggles, making it much safer for the teams, and the people they rescue.

TWENTY-SEVEN
OUR HELICOPTER CRASH AND NEW AIR RESCUE CENTRE

I LOVE WELLINGTON and could name a million reasons why I live here, but the weather is not one of them! A Wellingtonian's favourite saying is 'you can't beat it on a good day', and it's true; on a clear blue sunny day, with the water sparkling in the harbour, I don't want to be anywhere else.

Unfortunately, we must endure the bad days to get to the good ones. Americans think of Chicago as 'the windy city', but its mere 18-kilometre-per-hour average wind speed pales compared to our 29 kilometres per hour. Wellingtonians are so proud of our wind that we have signs and sculptures all over town embracing it. Children born here grow up thinking rain always falls sideways.

Wellington Airport, where Life Flight is based, has on average 175 days a year of gale-force winds (75 kilometres per hour or more). It's okay to fly in these conditions as it is well within aircraft limits, but start-up and shut down can pose an issue as the rotor blades could potentially hit the tail fin if they started flapping around in strong winds.

A decision was made to build a custom-designed facility at Wellington Airport. Once completed, the new hangar in the Air Rescue Centre (ARC) would be wide enough so we could start up inside and then pull the trailer out with the helicopter running. The fixed-wing hangar was designed to protect our patients from bad weather by loading and unloading them under cover, and then putting them straight into an ambulance.

Spirits were high as we looked forward to moving into the new ARC in February 2003. These high spirits were shattered when our helicopter crashed during a late-night call-out to Masterton on 14 January 2003.

RUTH ZEINERT, OUR crewperson on board, was calm, cool and collected when she called and told us that the helicopter had hit some trees in the Tararua Ranges, a mountain range which runs between the Wellington and Wairarapa districts. Everyone was okay but the helicopter was damaged and unable to land. Ruth, the doctor Michelle and the nurse Henny were all off the helicopter, but Sue, the pilot, was hovering above the airfield while the fire service cut away the skids so that she could make an emergency landing in a bed of tyres that had just been delivered to the airport.

Some of the waterfront sculptures showing Wellington's wind.

By the time we arrived at Masterton, aboard the fixed-wing air ambulance, Sue had managed to safely land the helicopter. She was suffering from a ligament injury to her hand, but other than that, no one on board had been injured. The helicopter was heavily damaged and had obviously struck trees with some force. In addition to the skids, one of the rear tail fins had been ripped off the helicopter.

When I looked inside the helicopter I was pleased to see that everything had been strapped down as it was supposed to be, so no missiles flew around the back of the helicopter to injure anyone.

The crash was serious enough that the helicopter never flew again but we were thankful that everyone on board was safe. Our operating company provided us with a backup helicopter the following day so we could keep providing services to our community.

All the people on board did a spectacular job that night and we learned a lot as an organisation. Helilink, our operating company, and Life Flight initiated many changes to procedures, and began requiring the medical team to wear helmets, something many of the medical staff had previously resisted because they regarded them as heavy and constricting.

When speaking about air crashes we often speak of the 'Swiss cheese' effect. Lots of little things going wrong until the holes all line

The damaged interior of the helicopter.

up, and then the accident occurs. That was the case with this crash. The astounding thing was that even after the helicopter hit the trees, Sue was able to keep flying it.

The helicopter crash put a damper on 2003, but having all four of the crew on board the helicopter alive and well for the ARC opening in March made it okay.

One of the good things to come out of the crash was a replacement BK117, which had fewer flying hours and more advanced radios and cockpit instruments than our old helicopter. The new helicopter arrived in December, allowing us to move into 2004 in our new ARC, grateful that no one had lost their lives.

The new Westpac Rescue Helicopter arrives at Wellington Airport, December 2003.

TWENTY-EIGHT
RURAL PEOPLE ARE TOUGHER THAN CITY FOLK!

KIWI FARMERS HAVE a reputation for being tough and stoic — a cheese scone and a cup of tea is enough to set them right. Over the years, I have found them to be tougher, more resilient, more grateful and have a better sense of humour than most of the city people I know.

As a boy who grew up in New York City I'm pleased to have had a chance to learn this. I will always be a city boy, but I sure hope that I have managed to adopt some of the characteristics of my rural friends.

One evening, we were called out to an Upper Hutt farm to attend a farmer who had spent 30 hours trapped under his tractor. His brother, who lived next door, became concerned that he hadn't seen or heard from him for more than a day so he started searching. His brother found him trapped under the tractor on the back of the property. After failing to free him, he raced back home and called for help.

The farmer, a man in his sixties, had been working on hilly terrain when he fell off the tractor and was pinned under the wheels. He was suffering from several injuries, was tired, cold and hungry, but wasn't so sure he needed to go to hospital. He was embarrassed about the situation and grateful that so many people had come out to rescue him.

As I kneeled by him, the first thing he said was: 'I'm so sorry to bother you guys, I'm sure you have better things to be doing.' It was not the first or last time I heard this sentiment from a farmer we rescued.

I find it amazing that someone living on a rural property in the middle of nowhere is noticed missing within 24 hours, but we often hear stories of people being found dead in their flats, months after they died, and not one of their neighbours noticed.

MANY PARTS OF the Greater Wellington Region and the top of the South Island have large areas of rural property. Large farms, lifestyle blocks, vineyards and fishing villages are spread throughout. The people who live and work on farms, or other remote areas, face many different challenges to city folk, especially when they are out on their land.

One of their main difficulties is the lack of cell phone coverage in rural areas, making it impossible to call for help. I always encourage farmers to carry a Personal Locator Beacon in their pocket in case they ever get hurt or trapped and can't make it back to their homestead to call for help. It could be a life saver for them.

The most dangerous piece of machinery in my flat is my blender, but farmers have all sorts of devices with the potential to cause major damage. When things go wrong, they can go very wrong.

Back in 1993, we attended an accident where a man had been digging holes using a post hole auger attached to his tractor. Somehow one of his trouser legs got caught up in the auger, and by the time someone managed to shut it down, his leg was partially amputated. When the first ambulance reached the farm, he had already used his belt as a tourniquet and was busy giving his farm-hand instructions about what needed to be done around the farm. I'm surprised he hadn't put the kettle on as well. We flew him to Hutt Hospital but unfortunately his leg couldn't be saved.

The remoteness of many farms often makes rescues difficult. I recall several jobs where an ambulance crew has made it to the homestead, only to find that the patient is kilometres inland, in an area only accessible by horseback, 4WD bikes, or helicopter. But rural communities are also supportive of one another, and by the time we arrived at the scene, the injured person was usually surrounded by friends and neighbours who came over to help.

Three- and four-wheel bikes have also caused more than their share of call-outs over the years. There were a couple of years when I am sure that I responded to more farm bike crashes than car crashes. I have been to gas bottle explosions, wood cutting accidents, hunters accidentally shot, people falling out of trees, and myriad incidents in our rural areas.

If that wasn't enough, animals contribute their fair share. Cows, bulls, horses, pigs and sheep have all been responsible for many serious injuries.

I have concluded that cattle do not like being loaded onto trucks, based on the number of them that have done serious harm to the

Helicopter, ambulance and fire service workers look after the victim of a quad bike accident.

workers trying to load them. One of these rescues was for a nearly 80-year-old man who was seriously injured while loading a bull onto a truck. The man was trampled and mauled by the bull, and according to one of the witnesses he was 'tossed around like a rag doll'.

After other workers managed to get the bull away from the injured man, the bull attacked another person, a farm vehicle and then a police car arriving at the scene. The bull was contained by the time we arrived so I didn't have to be worried about wearing a red flight suit, and being a red rag to a bull!

The paramedics on the scene were dealing with several major injuries. Aaron, our paramedic, had to sedate and intubate the older man using RSI, prior to taking off. On the flight to hospital I had a simple but important job, squeezing the bag mask attached to the tube every five seconds. While I did that, Aaron dealt with a host of other serious problems. Our pilot, Dean, knew that time was of the essence and managed to use a shortcut through a pass, saving us some precious time. Once in hospital our patient was stabilised in ED, sent for emergency surgery and then spent time in ICU.

As usual, it was a big team effort that helped the man survive the attack and be discharged several days later.

TWENTY-NINE
IMAGINE IT. DONE.
— PART 1

ONE COLD, BLUSTERY morning in October 2004, 12 steel-hulled 72-foot yachts slipped out of Portsmouth Harbour and into the English Channel to begin the 7-month Global Challenge yacht race. Dubbed 'The World's Toughest Yacht Race', with each yacht crewed by a professional skipper and 17 amateur sailors, it was considered the ultimate sailing challenge.

One of the things that made this race unique was that as the yachts are identical, winning came down to teamwork, strategy and some luck. The teams competed hard while sailing and then relaxed and partied together when they reached shore at each of the pit-stops.

The 2004 course would take the boats on a westerly route around the globe, sailing 54,000 kilometres with stops in Buenos Aires, Wellington, Sydney, Cape Town, Boston and La Rochelle before returning to Portsmouth, much of it through the perilous Southern Ocean.

Under normal circumstances that is daunting enough. The Southern Ocean would be better named the ocean of storms as, unencumbered by any land mass, wind and water tear around Antarctica like an out of control locomotive, generating ferocious storms and massive waves.

Furthermore, the westerly or 'wrong way' route meant the yachts not only battled the brutal prevailing winds and currents, but also hazards such as rogue waves, icebergs and fog. Southern Ocean races have a long history of disaster, including dismastings, overturned boats, smashed keels and rigging problems.

And that's if you're lucky; an additional nightmare for any round-the-world sailor is getting seriously sick or injured when they are days or weeks from medical help, especially in the Southern Ocean. Rescue to such a remote and inhospitable location is possible, but often means a long and hazardous wait until a ship can reach a distressed vessel or sailor.

On 28 November, after a couple of weeks on shore, the 12 teams left Buenos Aires on the second leg that would take them south around Cape Horn and on to Wellington. It was a gruelling leg that would take about six weeks and test the crews' physical and mental fortitude.

Generally, the crews alternated three- to four-hour shifts before heading below to rest for another four hours. On 22 December, nearly a month after leaving Argentina, one of the yachts, *Imagine It. Done.* (IID), was well into the leg when the boat fell off a rogue wave. The sudden drop caught one of the crew, John Masters, unaware and he lost his grip while moving about below, slamming into a bunk.

At first, he only complained of some minor rib pain and a large bruise on his ankle, and assured the crew he was okay. He continued crewing on the midday watch as usual, but during the 2000–2300 watch that night, it soon became apparent that John was not his usual jovial self.

The yacht's medic, David Roche, examined him and ordered him not to go on the 0200–0500 watch. As his condition deteriorated the doc arranged a 'hospital' bunk where he could administer the drugs to help John's recovery. By Christmas Day, he was clearly unwell, running a high fever and showing signs of infection.

All the boats had comprehensive medical kits and a designated medic — someone with some form of medical background, although that varied from doctors to medical students to veterinarians. Fortunately for John, David was a practising GP in the United Kingdom with years of experience. After consulting with the onshore medical team, he started John on intravenous antibiotics. As the days rolled on, though, John's condition worsened, and it was clear he needed urgent medical care.

The yacht's skipper, Dee Caffari, was in constant contact with the race organisers in the UK looking at their limited options. After considering all options, Dee, the race organisers, medical staff in the UK and Wellington Hospital agreed that IID would withdraw from the race and head, as fast as possible, directly to the Chatham Islands.

The Life Flight fixed-wing air ambulance would meet them on the island, as they routinely flew patients the 750 kilometres to Wellington for any medical care that was not available locally. Even though it would take nearly a week, it meant that John could get to Wellington Hospital at least two or three days sooner than if the yacht continued directly to Wellington.

The helicopter wasn't initially considered as it only flew to the Chathams as a fuel stop, before it performed winch rescues off a ship or yacht further out at sea. Patients were then taken back to the Chathams and put on the fixed-wing air ambulance to be flown to Wellington.

On 26 December, Life Flight was given a heads-up that they would be needed sometime around New Year's Day, depending on the progress the yacht made. We followed them on the Global Challenge website which gave six-hourly updates on each yacht's position, as well as their current speed and track.

Meanwhile, on board the yacht, David, Dee and the crew were doing everything they could to keep John alive.

All the yacht crews kept in contact with each other via radio, satellite phone or email throughout the race — they understood how dire the situation was and offered moral support because there was little else they could do.

But the clock was ticking relentlessly. The medical supplies on IID were running low with more desperately needed. Without hesitation, the two nearest yachts abandoned the race and set course to rendezvous with IID, literally in the middle of nowhere, to transfer them the much-needed supplies. The stakes had changed dramatically; everyone in the Global Challenge now knew that the race was to keep John alive and were willing to do whatever it took to ensure that happened.

By New Year's Eve, they were still at least two days from the Chathams, but John was now so ill that Dee and David were forced to consider the unthinkable — how and where they would keep his body if he died.

After consultation between David and the medical teams in the United Kingdom and Wellington, it was clear that a helicopter rescue — winching John off the yacht and bringing him back to the Chathams — was now his best chance of survival.

I was the duty helicopter crewman when the call came in from RCCNZ late that afternoon requesting the rescue, somewhere southeast of the Chathams, sometime over the next few days. Grant Withers, our duty pilot, checked the weather forecast for Wellington and the Chatham Islands. Luck was on our side; the weather would be suitable for the next two or three days.

After advising RCCNZ that we would accept the mission, Grant and I checked the latest yacht positions and as IID wouldn't be in range until late the next day or possibly the day after that, we wouldn't need to leave till tomorrow, New Year's Day. This gave us more time to prepare, which was good, because flying beyond the Chatham Islands is a complex, expensive, and potentially risky operation which needed careful planning and coordination. Although it was routine for the fixed-wing, in my 25 years at Life Flight the helicopter had only performed five of these missions. We had a long checklist of things to do and arrange before the following morning. Even though I was not scheduled to work the night shift I was not going to leave Grant and Jon Leach, the helicopter crewman, working New Year's Eve, to do all the work. We were a team.

As well as the helicopter and the fixed-wing plane we would also need a smaller non-pressurised 'spotter' plane which would keep a watchful eye on the helicopter during our hours flying over water.

About 700 kilometres of the trip to the Chathams were over water and would take three and a half to four and a half hours, depending on wind speed and direction. As we could only carry two and a half hours of fuel in the tanks, we would have to take two to three extra hours' worth of fuel in the helicopter's cabin.

I thought I might briefly join a group of friends for our planned New Year's celebrations, but then we hit a major obstacle and I knew that was not going to happen.

The helicopter would need to refuel twice at the Chatham Islands; once when we first arrived to have enough fuel to reach the yacht and then again for the flight back to Wellington. The spotter plane would also need to refuel before heading back to Wellington. A phone call to Air Chathams, who organised fuel for us on the island, brought bad news. There was not enough Jet-A1 fuel on site for our needs.

Without fuel on the island, we couldn't leave Wellington. We worked with RCCNZ to come up with a solution, and after much discussion hit on a plan. RCCNZ organised Air Chathams to send their Convair plane from the island to Wellington early the next morning, carrying empty Jet-A1 fuel drums in their cabin. We would fill the drums and they would fly them back to the island, ready for us to use when we arrived.

But this meant the costs for the rescue were mounting up, flying the fuel to the Chathams was more than $40,000 on its own, but the cost of not getting the fuel was higher — John would probably die within the next few days.

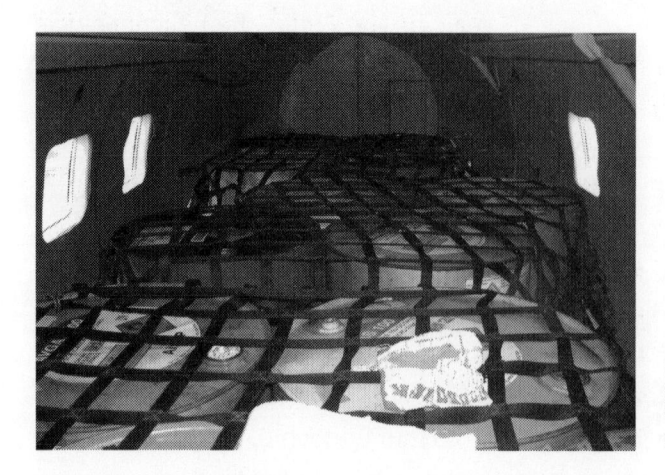

Fuel drums loaded in the back of the Air Chathams Convair plane.

Jon, Grant and I worked through the evening, gathering and testing the equipment and finalising our plans. We would be good to go anytime from sunrise, but after checking IID's position, we saw the yacht was making agonisingly slow progress due to the weather and wouldn't be within our range until later in the day, at best.

We called the yacht and they confirmed that neither the weather nor John had improved. It was frustrating, but there was nothing we could do, so around 11pm we called it a night. I briefly contemplated trying to meet up with my friends for that New Year's drink but I was tired, hungry and smelly so I decided to spare them and headed home.

The next morning, Wellington Free Ambulance assigned Iain MacKay, an experienced winch-trained paramedic, to the helicopter for the mission. Iain and I had worked together many times, and I hope he felt as confident having me as the winch operator as I did having him as the paramedic.

Iain met Grant and me at the hangar just after 8am and we sat down to have a thorough briefing and planning session. Since it was only a week or so past the longest day, we had plenty of daylight in which to perform the rescue. Thankfully, the yacht had sped up and it looked like we might have a chance later that evening.

We studied pictures of the yacht intensely to try and get a clear idea of what lay ahead. It was a mass of ropes and steel cables with a very tall 30-metre (100-foot) mast — nearly as high as a ten-storey building. It made the *Terminator* rescue, with its 15-metre mast, seem like a piece of cake. We would need to stay well clear, although not being in gale-force winds this time round was a help. We decided the best option was to winch Iain down to the aft starboard deck — the back right-hand corner of the yacht.

With the information to hand, we decided on an 11am departure from Wellington, meaning that we could get out there, refuel and be ready to head out to the yacht by 5pm. While we were busy with our planning, the Convair arrived and Phil, our daytime fixed-wing crew, gave them a hand to fill the fuel drums in the back of the plane. Once they headed back to the Chathams, it was a relief to know that a huge problem was solved.

Our friends and neighbours at Wellington Airport, Vincent Aviation, provided a Cessna 402 as the 'spotter plane' for the mission. It would fulfil many tasks, mainly by carrying the paramedic, the winch and all the medical equipment which needed to be removed from the helicopter. Since the plane was faster than the helicopter, it would depart later and fly higher than us to keep in radio communications with the helicopter, and relay radio messages between us and air traffic control. During the rescue phase, they would leave before us, locate the yacht and relay information about the weather, sea state, etc. back to us.

The plane also supports the helicopter if it runs into trouble and has to ditch into the sea. As the spotter plane is non-pressurised, it can open its door and drop supplies to us if necessary. My mum once asked me what the spotter plane was for, and I told her that they were there so someone could snap a picture of where I died if we crashed into the sea. I'm still not sure if she knew I was joking.

As the Chatham Islands time zone is 45 minutes ahead of Wellington, Grant determined that if the rescue was to be completed that day, the yacht needed to be within 315 kilometres of the Chatham Islands by 6.15pm, Chathams time. This would give us about 90 minutes to fly to the yacht, 45 minutes to perform the rescue, and another 90 minutes to get back to the Chathams before we ran out of daylight.

Unfortunately, when we rechecked its position, we saw IID had slowed and it seemed unlikely that the rescue would be carried out that day, but it was hard to know for sure. Even if we couldn't get out to the yacht until the next morning, we still had to get to the Chathams that day.

My first helicopter trip to the Chathams 11 years earlier had been a thrilling adventure full of unknowns, but this time it was simply a means to an end. About 30 minutes after take-off, we passed the last bit of the North Island. Next stop, Tuuta Airport.

Fuel management was vital on this leg, and for Grant this meant varying our altitude, trying to take advantage of the different wind speeds and directions at different altitudes. For me, this meant manning the pump to move fuel from the drums.

Soon after we emptied our last reserve fuel into the main tank the fuel gauge began going down — we now had two and a half hours of fuel left. According to the GPS, we were only about 110 kilometres, or 30 minutes, from the Chathams. About 15 minutes later we saw land in the distance, and even though we knew we were on track it was still a relief to see terra firma. We were almost there!

After nearly four hours we landed. I called IID and they were still about 390 kilometres from us, about 75 kilometres beyond our line in the sand. They only had two hours to cover that distance so, disappointingly, it was not going to happen that evening.

Nevertheless, we still had to set up for the rescue. We removed the empty fuel drums from the helicopter then filled the fuel tanks. The spotter plane wouldn't need fuel until after the rescue. We then moved all the rescue and medical equipment from the spotter plane to the helicopter. Barely an hour after landing, we were all set to go.

I called the yacht with the faint hope that they had sped up, but they were still out of range, and there was no way they could cover that distance in the next hour. It was agonising for us — to be so close but not close enough. We would have to wait until dawn, delaying John's rescue and transport back to Wellington. For him to survive so long, only to succumb to his injuries overnight, especially since they were so close, would be unbearable, and the temptation to set a new line in the sand and go now was high. After all, they would only be 40 kilometres or so beyond the original line by the time we reached them.

However, as frustrating as it was, the line in the sand was 315 kilometres for a reason — this gave us enough time to safely get to the yacht, perform the rescue, and return to the island by dark. There was not enough fuel on the island for us to attempt the rescue twice, and we needed a safety margin in case we hit stronger winds or if it took longer to do the rescue than expected. Every kilometre past the line decreased our safety margin and it was an unacceptable risk.

Calling the yacht and telling them that we would have to wait until morning was one of the most heart-breaking phone calls I have ever had to make. Their disappointment was clear, but the news was not unexpected. They had managed to keep John alive for over a week already so they were hopeful he would survive another night. We all were.

THIRTY
IMAGINE IT. DONE.
— PART 2

WE HEADED TO the Hotel Chatham where we were staying that night. When we met for dinner, it was evident that the staff and guests had had a good night celebrating the arrival of 2005. Most of the people seemed to be moving slowly, possibly still nursing hangovers, but we were greeted warmly.

News travels fast in such a small community, so everyone in the hotel knew about the upcoming rescue. People chatted with us, many offering to buy us drinks, which we declined. Although we enjoyed the local blue cod and crayfish for dinner, we were all distracted, worried that John wouldn't last the night. It was a sobering thought.

We had a great meal and mingled with the others at the hotel before heading off to try and get a good night's sleep to be ready to take off by sunrise at 5.50am.

I was up before dawn and called the yacht — it was great news — they were now about 160 kilometres from us, and John was alive and no worse than yesterday. It was a huge relief.

Over breakfast, we had a mission briefing with the spotter team. The helicopter would take off first and head to the yacht's position; the plane would take off shortly after and then make its way to the yacht — they would be able to confirm the yacht's position and conditions in the area for us.

I called back to base in Wellington and told them the plan; they would need to get the air ambulance on the way to the Chathams by 7.00am — the trip for the Metro was under two hours each way, but given the time zone difference, we expected to be back with John between 8.30 and 9.00am local time.

We gathered at the airport, did our safety and equipment checks, got back into our immersion suits, and just before dawn we headed towards the yacht — the rescue was on.

Shortly after take-off, we established comms with the yacht on the marine radio. The crew member was overjoyed when they heard we were on our way. The weather conditions were just okay, about 20-knot winds, with waves of around 2 metres and a lot of low cloud.

About 30 minutes later, the spotter plane reported they had the yacht in sight and that the clouds were not a problem, at least for now.

We had previously emailed the IID crew instructions about the helicopter rescue. Dee, the skipper, knew that she would pick a track,

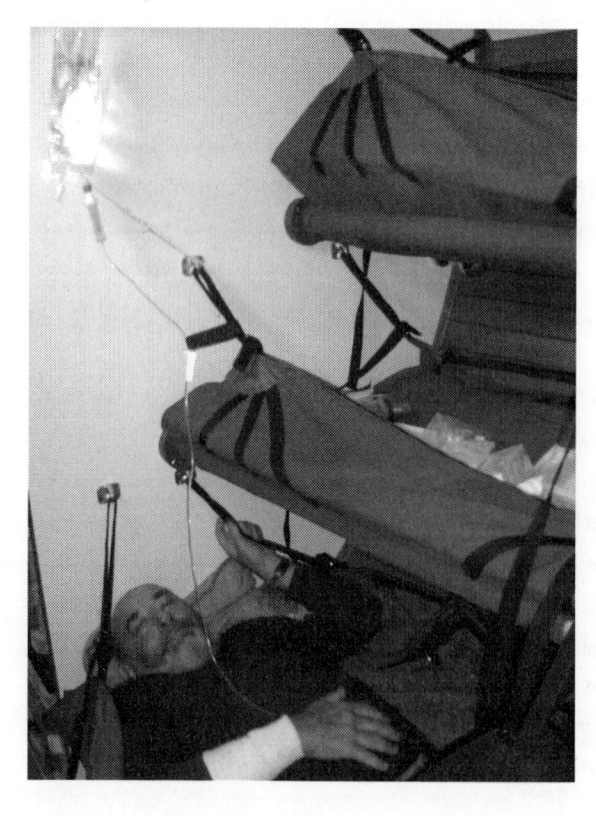

John Masters in his bunk.

almost directly into the wind, and keep on that track no matter what the helicopter did. The worst thing she could do, from our perspective, was 'try to help' by steering the yacht to where she thought we would want it.

As we approached, we saw a lot of people on deck. We noticed that the crew was looking extremely competent — the yacht was being held directly on track, all the obstructions had been removed from the winching area, and there was even someone in a wetsuit and fins, ready to jump into the water if anything went wrong.

I opened my side door and sat in the doorway as we circled the yacht and confirmed that everything was as it had looked in the pictures. Most of the crew were waving and smiling at us and I could almost feel their relief as we flew around them.

Our plan still looked good so Grant moved in towards the yacht and I threw the weighted sandbag on the end of the hi-line onto the deck. Once the sailors had hold of it, Grant backed away, positioning the helicopter so he could still see the yacht.

The biggest danger for us was the 30-metre mast swinging about wildly as the yacht moved through the 2-metre waves. While Iain got ready, Grant and I discussed what we would do with the mast moving so unpredictably. However, the yacht keeping so well to the required track, meant Grant could keep the helicopter's nose directly into the wind, while giving me enough room to operate off the aft deck.

Once Iain was ready, I did a safety check of his harness and equipment and then moved him to the door. Attached to the winch hook was a huge sack full of milk, fresh fruit and newspapers — now that the IID crew had withdrawn from this leg of the race, the least we could do was give them a few creature comforts to keep their spirits up until they reached Wellington.

I put Iain out the door on the winch and lowered him towards the sea below. Grant still had the yacht in sight so stayed in relative position to them. The sailors aboard were doing what we had requested — pulling on the hi-line, keeping it taut, and ensuring that Iain didn't swing about wildly below the helicopter.

When he was low enough, I started talking Grant in towards the yacht. As soon as he moved forward he lost sight of it, instantly making both of our jobs much harder. As the yacht pounded through the waves the mast was moving dramatically, almost in a square shape. When the yacht fell off a wave, the top of the mast moved forward and as it climbed up the next wave it moved backwards – at the same time the yacht was rolling left and right. The mast seemed like a mean-spirited creature waving a lance about trying to stop us from carrying out our mission. Grant had very little to reference but concentrated on keeping well away, as if we struck the mast, it could cause us to crash into the frigid water below.

We were struggling to get close enough to the yacht to get Iain aboard. The safest approach was from behind to ensure that Iain didn't get tangled up in the mast, the ropes and steel cables around it. This wasn't working so it was time to come up with plan B.

We decided to abort the winch attempt and Grant moved the helicopter back into a position where he could see the boat. I winched Iain back to the helicopter. He suggested getting him close to the steel cabling so that he could work his way down to the deck but we ruled that out as too dangerous. We all decided that we would give it another go, still from behind the yacht, but we would have to work quickly to get him on board, off the hook, and then move the helicopter back away from the swinging mast. However, we would have to drop him closer to the water before moving forward, so Grant had as much reference time as possible.

I called them on the radio and said we were about to start again. Their relief was palpable. I found out later that when we pulled Iain back into the helicopter they had assumed that we couldn't get him aboard and we were aborting the mission altogether. They were sure we were going to leave John behind, condemned to an almost certain death.

We started again and I got Iain just above the water before I called Grant to move in towards the yacht. Between the waves, the helicopter moving about in the wind and the movement of the yacht, Iain ended up going for a bit of a dunk in the water as we moved forward — it wasn't the first time I had dunked him during a rescue and it probably wouldn't be the last!

The crew pulled like mad on the hi-line, and as Iain got close they grabbed him as I put out a few extra metres of winch cable to ensure that we didn't pull him off the boat unexpectedly. As soon as he had his balance he removed the winch hook from his carabiner and held the hook out — our signal that he was clear and Grant could move back to his reference point.

Those had been some of the tensest moments I have had as a winch operator. Everything seemed to be conspiring against us, but we successfully managed to get Iain aboard, albeit a bit wet.

I watched as John crawled towards Iain on his hands and knees. They got John into the nappy harness and once he was secured, the crew helped John to the rear corner of the deck. I could see, even from a couple of hundred feet away, that John was in a lot of pain but at least he could walk with assistance and didn't need to be carried.

Dave and Iain assist John out of the helicopter and onto a waiting stretcher.

Once they reached the winching spot, I let out winch cable so that they could use the hi-line to pull the winch hook back towards them. When Iain had it in his hands, he put the carabiner from John's harness on the hook, followed by his own. I watched him give a final safety check, and when he was satisfied that everything was okay, he looked up and gave me a big thumbs-up.

Winching off the yacht is much easier than winching onto it. Grant moved the helicopter in towards the yacht, but well above the height of the mast. We performed a 'snatch lift', meaning that we used the helicopter to 'snatch' Iain and John up and off the deck, and then quickly moved away from the yacht and the swinging mast to a position where Grant could see the yacht. As soon as they were clear, I winched them in towards the helicopter.

As we sat John in the helicopter doorway he gave a big wave and his own thumbs-up to his waving shipmates below. He was relieved and happy to be off the yacht and on his way to the hospital. We retrieved the hi-line, gave a final wave to the yacht and headed back to the Chatham Islands. The radio message from the yacht was clear — they were grateful to us, and happy that John was on his way to much-needed hospital care.

John was clearly unwell but there was not much that we could do for him in the helicopter. He needed to get to Wellington Hospital for urgent surgery — nothing else was going to help him at this point.

He was very thankful that so many people had gone to so much effort to save his life. It turned out that he was a Kiwi who was living in London and was disappointed that he wouldn't get the opportunity to sail into Wellington. He also knew that despite the tremendous amount of money spent, the time training and then weeks aboard the yacht, his dream of circumnavigating the earth was over. It was a sad realisation, but he was extremely grateful to be alive.

Our flight back to the Chathams took about 45 minutes and we had word that the air ambulance was on its way with a flight doctor and nurse. We also had a big surprise for John — his wife Lorraine, who had recently arrived in Wellington from the United Kingdom, was aboard the plane.

After landing, we helped John out of the helicopter and Iain and I supported him as he wobbled his way towards a waiting stretcher. Not only was he suffering from sea legs, but for the better part of the last ten days he had been lying in a bunk, in a yacht, rolling about in high seas. We got him as comfortable as possible while we waited for the arrival of his ride to Wellington.

One of the nicest moments of my career was watching the reunion of John and Lorraine. The love they shared and their joy at being reunited was evident. They hadn't talked to each other since the ordeal began and neither had been sure that they would ever see the other one again.

The medical team from Wellington gave John a quick check, and then we loaded him onto the Metro. Since the Metro is a much bigger plane than the Cessna spotter plane, we sent Iain and most of the rescue and medical equipment back with them.

After they took off, it was time to get the helicopter and spotter plane refuelled and head home ourselves. Little did I know my IID adventure was far from over!

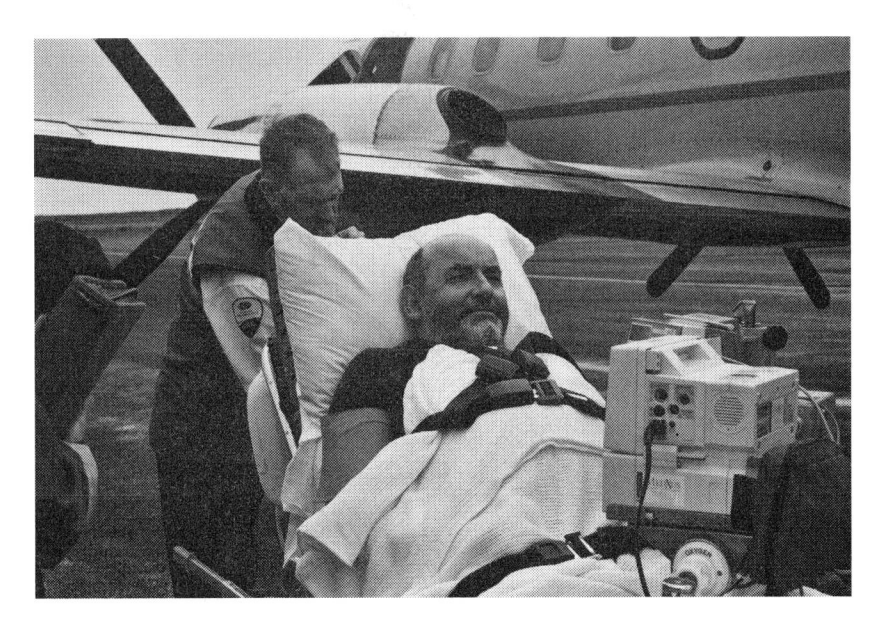

Iain MacKay helping load John onto the air ambulance.

The trip back to Wellington was a slow one. We were flying into strong headwinds which reduced our speed, however everything was going as planned, until I ran into a bit of a problem with a fuel drum.

I went to open the second drum and found that its cap was well and truly jammed in place. I checked the other drum, and its cap opened so I pumped the fuel from it into the auxiliary tank.

It turns out that if you ever want to make a pilot snap his head around quickly, all you need to do is be hundreds of kilometres from land and say, with a bit of concern in your voice: 'Do you have a sec?'

I explained the problem and said I was sure I could solve it with some banging and brute force. Grant told me that we were approaching our point of no return so I had five minutes to get the bloody drum opened or we were turning around and heading back to the Chathams.

That was all the incentive I needed! I was not worried about our safety because I knew we had enough fuel to get back to the Chathams but I wanted to get home. I worked out that the steel drum expanded a bit when we gained altitude, jamming the cover. After a minute or so of struggling with it, I got it open.

The trip home took over four hours and by the time we got back, John was already in surgery. When they cut him open, they found a massive infection. The doctors told John that it was unlikely he would have survived another 12 hours without surgery.

It was gratifying to learn that using the helicopter instead of waiting until the yacht reached the island had saved John's life. We were extremely glad that waiting until the next morning didn't kill him, yet we were equally sure we made the right decision in waiting. Many aviation accidents are caused by pushing the boundaries too far — Grant had pre-determined how far we would go out, and it was important we stuck with it. It is comforting flying with experienced pilots who make excellent decisions!

THIRTY-ONE
RESCUE DAVE

IID ARRIVED INTO Wellington Harbour three days later, on 5 January. We surprised the team by doing a fly-by of the yacht as it sailed up the harbour entrance, and we surprised Dee even more when she spotted her partner Harry waving to her from the helicopter.

Iain and I were on the dock as the crew came off the yacht and there were hugs all around. It was a fun thing to be part of. A few days after that Unisys, IID's main sponsor, held a thank-you dinner for everyone involved in the rescue. It was lovely to finally meet and celebrate with the team.

A few of the guys jokingly suggested that I should take John's place on the next leg of the race, from Wellington to Sydney. I am a rescue helicopter crewman, not a sailor, although I had attended a sailing course as a way to meet a woman a few years before, but I had never sailed outside of Wellington Harbour and really had no interest in doing so. However, I was merry and full of bravado, so I said of course I would go — thinking that there was no way in the world that the race organisers would ever allow anyone who had no training or experience to get on the yacht.

I spoke to Dee that night and she thought it was a great idea for me to come along. She would teach me to sail, she said, and I had already earned my stripes and was welcome aboard. I told her that if the race organisers said yes, I'd be there.

Much to my surprise and horror, the race organisers agreed! I was officially a 'legger' (a person taking part in only one leg of the race) for Leg 3 of the Global Challenge. I joined the IID team for a planning session a few days later. There were four 'Daves' aboard the yacht for this leg of the race and we all ended up with nicknames — I was given the name Rescue Dave, which seemed a good fit. I liked it so much I took it on as my new persona and email address.

I had a few training sails around the harbour and then a few weeks later, on 6 February, I found myself aboard the yacht heading towards the starting line. I looked up with pride as the rescue helicopter circled us with a surprise guest, John Masters, waving us off.

It took us seven days to make the Tasman Sea crossing and they were seven great days. I loved hearing the stories of the first two legs of the race, particularly about sailing through the Southern Ocean in waves that were so large they would describe how big a house they could build in the trough. Hearing about John getting hurt, the fear he wouldn't survive, and the rescue from their perspective was amazing. Watching from the deck, they thought the mast was going to strike the helicopter several times and were relieved when it didn't.

My favourite part of the leg was the night when we got caught in a storm. I was loving the adventure of it all. Dee did two things during the storm which still make me smile. A shift turnover was coming

Dave and crewmates on deck in the middle of the Tasman.

up at around 2am, and when Dee spotted some of the other leggers coming up onto deck she shouted out 'no fucking leggers on deck'. I looked at her and using hand signals asked if she wanted me to go downstairs. I got a firm shake of her head in response, I was okay to stay up top.

A few minutes later we were battered by immense waves and I ducked down to take cover in a section called the keyboard, in the middle of the boat. To my amazement, a few seconds later Dee shouted out, 'Is Rescue Dave still with us?' I got a huge Dee smile when I popped my head up and gave her a wave. I could not believe that with the chaos going on around her, she still managed to keep an eye on everyone across a 22-metre (72-foot) deck.

After being near the middle of the pack for a good part of the leg, we ended up making a poor strategic decision and entered Sydney in last place. As disappointing as that was, it was a memorable arrival into Darling Harbour. It was a beautiful night and there was hardly a breath of wind as we sailed up the harbour carried by the incoming

Dave and crewmate Darren on the way to Sydney.

tide as much as the wind. We passed the Opera House and then crossed the finish line somewhere near the Sydney Harbour Bridge.

The strange thing about my experience was, given there were two different teams aboard, working opposite shifts, I got to know half the IID team well but barely got to know the other half of the team at all.

Reaching Sydney was a milestone and I decided that I have a Southern Ocean crossing inside of me. Unfortunately, the 2004/5 event ended up being the last Global Challenge race run and I never got my opportunity to complete that goal.

I spent several days with the team in Sydney before saying my goodbyes and heading home to Wellington. It was sad to leave but I knew that I would stay in touch with at least some of the team.

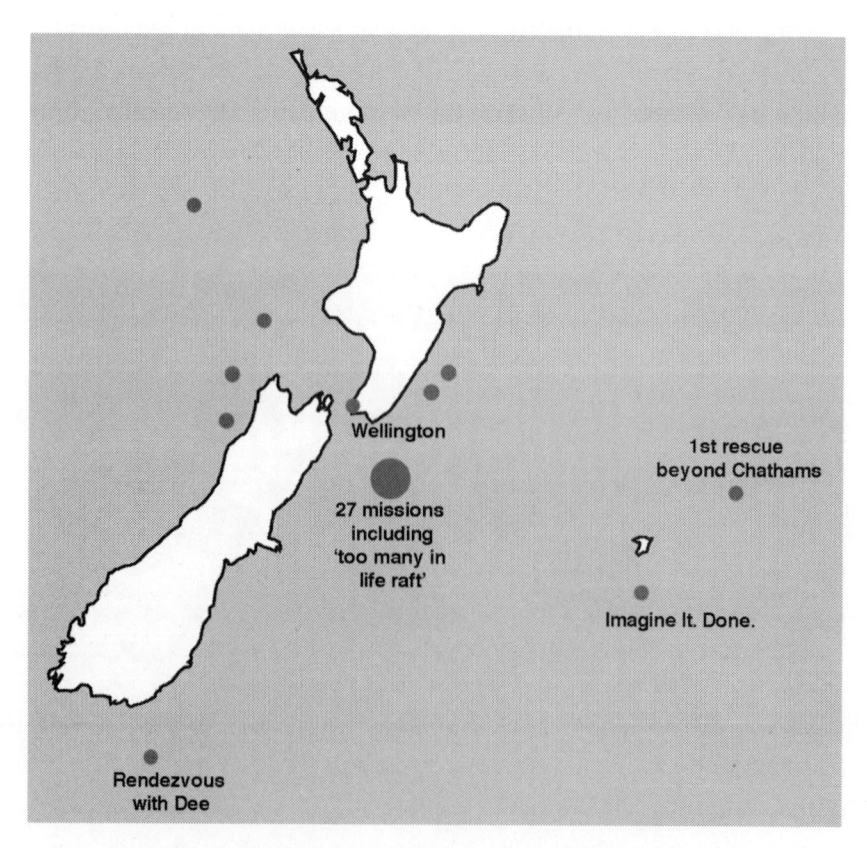

A map showing some of the locations of offshore rescues Dave was involved in.

190 Emergency Response

THE IID TEAM completed their round-the-world trip in July, nine months after they began. Soon after the race finished I heard that Dee had accepted a challenge from Sir Chay Blyth, the first man to sail solo the wrong way around the world, and would attempt to become the first woman to complete the same feat.

In October 2005, I flew to the United Kingdom and joined many of the IID crew to farewell Dee as she set off on her solo attempt. I followed her tracker and blog daily and was excited for her as she overcame many daily challenges. It was hard to believe that a single person, man or woman, could sail the same yacht that needed eight or nine people on a normal voyage.

In January 2006, I got a call from her organising team. They needed a way to rendezvous with Dee's yacht as she sailed south of New Zealand, to collect the hours and hours of video footage she had recorded. I jokingly said I would head out by helicopter, but they were looking for a cheaper solution. As it happened, a helicopter ended up being the best way and on Valentine's Day I found myself aboard the Otago Rescue Helicopter, heading 160 kilometres out to sea, to collect a waterproof container with her tapes inside. The radio conversation I had with Dee that day was one of the most special I have ever had with anyone.

It had been nearly four months since she had seen another person up close, and here we were close enough to look into each other's eyes, but too far to give her a much-needed hug. She was so grateful that we were there that she gave the four of us in the helicopter a bit of a dance on deck. We stayed overhead for a little while waving back and forth, and I had tears in my eyes and both of our voices cracked when we said our last goodbyes as the helicopter moved out of radio range.

In January 2008, three years to the day after our paths first crossed, I had the honour of breaking a bottle of champagne over Dee's custom racing yacht which was built in Wellington. In November that year, I went to Les Sables-d'Olonne, France, to see Dee start the Vendée Globe yacht race. Her sponsors asked me to organise a plane to fly past and get some action photos of Dee when she sailed through the Southern Ocean again. On 29 December, I found myself on a plane near the Auckland Islands, 500 kilometres south of Invercargill, giving her a wave.

To date, Dee has sailed around the world five times. She is the first woman to have sailed single-handed and non-stop around the world in both directions and the only woman to have sailed non-stop around the world three times. She was awarded an MBE in 2007.

I have managed to catch up with Dee a couple more times as she took part in a round-the-world race which had a pit-stop in New Zealand.

In my wildest dreams, I would never have imagined that my rescue helicopter job would lead to so many different adventures and so many lifelong friends.

Dave and Dee Caffari on board *Imagine It. Done.*

THIRTY-TWO
ANZAC DAY — PART 1: CONFLICTING MISSIONS

IT WAS PURE chance that I was working on the morning of Anzac Day 2010. I had picked up an overtime shift due to start at 6am. At 5.50am, Colin Larsen, the crewman heading off duty, rang to tell me that I was needed for a 6.45am flight to Blenheim.

The helicopter had been requested to attend a road crash near Picton around 10.30pm the night before, but low cloud across Cook Strait had prevented them flying to the scene. Around 5.40am Lois, the duty flight nurse, rang to say that a critically injured man from the accident was now in Wairau Hospital ICU and needed to be urgently transferred to Wellington ICU.

Harry, the duty pilot, checked the weather and found that there was still some patchy low cloud in the area so the flight would have to wait until daylight at about 6.45am. Colin and Lois discussed using the fixed-wing, instead of the helicopter, but since it took an hour for the plane to get airborne, it would still be quicker to use the helicopter.

As I left my home which overlooked the airport, I watched an air force Iroquois helicopter land at the Defence Hangar, which was a short distance from our base at the airport. I assumed it was in Wellington to take part in the Anzac Day ceremonies, which traditionally started at dawn.

Once at the base, Harry and I prepared for the short flight to

Wellington Hospital to pick up the medical team and then make the 25-minute hop across to Blenheim. It was now daylight, so any scattered cloud across Cook Strait was no longer a problem.

Seconds after take-off, RCCNZ rang and asked if we could assist in locating the source of an emergency beacon that had been activated in the Pukerua Bay area, about 40 kilometres north of Wellington. The initial satellite pass had indicated the signal was coming from about 500 metres off the coast. Worryingly, an air force Iroquois helicopter had been reported missing on a flight between Ohakea Base, the home of the squadron, and Wellington.

Most beacon searches turn out to be false alarms due to inadvertent activations or hard landings. We had both seen the Iroquois at the airport, and the chance of one of their helicopters being in trouble seemed slim. Surely if one was missing, they'd be out searching and not sitting on the tarmac in Wellington?

We had to decide whether to abandon the flight to Blenheim to pick up a critically injured patient for something that could well turn out to be a false alarm.

One of the most difficult things we often face is making 'life and death' decisions without all the facts. They must be made quickly, based on the information available at the time, and using all our education, training and experience. And then we hope like hell we made the right decision.

There is the added pressure that the wrong choice could lead to someone's death and that anyone from a courtroom jury to an armchair critic would have the benefit of hindsight to judge our decisions.

If we chose to go to Pukerua Bay for the beacon search and the patient in Blenheim died while waiting for the plane, we would have to live with that decision. If we declined to go and there was a crashed helicopter with survivors that died because of a delay in being found, we would also have to live with that.

If I was ever involved in a helicopter crash, I know what choice I would want a rescue team to make. Equally, if I was the parent of a critically ill teenager who was anxiously awaiting the arrival of a medical team from Wellington, I know I'd want a different decision.

No matter what we decided, we would have to delay either a potential rescue or a hospital transfer. Fortunately, in this case we had another option; the fixed-wing air ambulance could head to Blenheim, and as we were the nearest rescue helicopter to Pukerua Bay, we concluded that the beacon search was the higher priority for the helicopter.

We agreed we had to accept the beacon search. The best solution we came up with was to offer the medical team waiting at Wellington Hospital the choice of coming with us to Pukerua Bay or staying in Wellington and waiting for the fixed-wing. If the beacon turned out to be a false activation, we could head directly to Blenheim from Pukerua Bay, meaning we would only delay the transfer by 25 or 30 minutes. However, if we took them with us, and we ended up involved in a full search and rescue, they would have to return to Wellington Airport by car, and the delay would be upwards of an hour.

As long as we ensured the doctor and nurse were not exposed to any unacceptable risk, taking them with us seemed the better option to us.

We advised RCCNZ that we would accept the mission, but due to the low cloud we'd have to take a low-level route, around the South Coast and up the West Coast, which would take twice as long — about 20 minutes.

As soon as we landed on the hospital helipad, I told Lois and the doctor, Lesley, what was happening, and what their options were. They chose to come with us to Pukerua Bay.

As we lifted off, I updated ambulance communications that we were responding to a possible helicopter accident at Pukerua Bay instead of the hospital transfer and requested that ambulances and a winch-trained paramedic head to the area.

One thing I have learned to trust over the years is my gut feeling. And my gut told me that the morning was quickly turning to crap. Even though the chances were still high that we would be going to Blenheim ourselves, I called Steve Reeve, the fixed-wing duty crewman, and quickly explained what was going on and asked him to get the fixed-wing pilots to respond to base just in case the plane had to go to Blenheim.

We learned from RCCNZ that the three Iroquois had left Ohakea before daylight and flown together towards Wellington when they got caught up in low cloud approaching Pukerua Bay. They had split up and one flew to Wellington Airport, one diverted to Paraparaumu Airport but the third helicopter was missing and no one could get in touch with them.

Our experience told us that it was increasingly likely they had crashed. If they landed safely and their radios were inoperable, someone on board would have called on their cell phone to report their position.

It was a horrible prospect, especially as there was a good chance that Harry and I could know the crew members. The year before, we both attended a joint air force and civilian helicopter winch conference and workshop. During this, we had met many air force pilots and crew.

Harry and I prepared for the beacon search as we flew along the coast, which still had low-level cloud hanging about. This required activating the directional finding (DF) unit and setting up the helicopter's radios to make them work with the unit. The DF used aerials mounted on the helicopter to detect and track an emergency signal put out by an Emergency Locator Transmitter (ELT), which is mounted to a plane, or a Personal Locator Beacon (PLB), which is carried by a person.

The change of our mission type also meant we had to change our frame of minds. While any mission needs the pilot to fly safely, flying around trying to track a beacon is different from transiting between two known helipads. The jobs I perform on a search are also different on each mission. It would be like heading into a supermarket to grab a few items for dinner and then getting a text saying you needed to get enough food for a dinner party for 10 people without having had time to plan. They are both shopping, but one is much more complex than the other.

As we approached the location, we picked up an intermittent and weak beacon signal which was a bad sign. If the helicopter was sitting on the ground intact, we should have picked up a strong signal. A signal that was detectable by satellite but not us meant either the helicopter was in the hills or their ELT antenna had been compromised — or both.

As we got closer, RCCNZ told us that the Iroquois which had landed in Paraparaumu was now searching the Pukerua Bay area. Two helicopters flying in close range with low cloud about meant we had to stay vigilant.

At this point, we decided that as this was now an SAR operation, we would have to drop off Lois and Lesley as we couldn't have any unnecessary people on board. I told them we were going to Plan B and would get them back to the airport in a police car. I called the police and asked them to have a police car meet us at a parking area at the north end of Pukerua Bay. I also made a quick call to Steve to let him know that the plane would be making the trip and that the medical team was on its way.

As we rounded the last big hill before the landing area, the beacon signal strengthened — it was coming from the hills to the east. The Iroquois crew confirmed that they thought the signal was coming from a nearby ravine. Harry and I didn't say anything, but the look we exchanged said it all — this was not an inadvertent activation, we were searching for a crashed helicopter and crew.

THIRTY-THREE
ANZAC DAY — PART 2: THE SEARCH AND RESCUE

OUR GOAL NOW was to find a way into the cloud-covered ravine where the beacon signal was coming from. We desperately searched as we flew over and around the area, and we watched the air force crew doing the same. As we flew overhead we could hear one and occasionally two beacon signals wailing below, a constant reminder of what was at stake.

The cloud still lingered and a blustery wind was blowing in from the sea, which meant Harry would have to contend with a strong tailwind if we entered the ravine from the western side. Ideally, we would enter from the east so our nose pointed into the wind, but as that was higher up the hill it was unlikely that the cloud would clear from that end first.

A few minutes later, we noticed that the cloud was slowly lifting at the seaward entrance to the ravine. Harry was confident that if it lifted enough for us to enter he could deal with the tailwind.

We kept an eye on the entrance and were relieved as the gap slowly widened. We discussed the potential risks of heading into the cloud-covered ravine and how we could mitigate them. As easily as it lifted, it could also drop again while we were in there. On a normal day, we would never consider entering the ravine in these weather conditions, but this was not a normal day.

We had a good idea of the terrain in the area thanks to our moving map and the fact we had flown over or past it hundreds of times before. There was a hillside covered with trees that rose towards a small peak as it headed inland.

We came up with a solid plan. We would enter the ravine and Harry would make sure we flew along as slowly as possible, given the tailwind. We knew that the cloud was a relatively narrow band of low-level sea fog and from where we were we could see the clear air above it. Once we entered, if everything went to crap we had an escape plan: Harry would fly straight up through the cloud layer and into the clear air. The risk was greater than normal, but it was manageable.

I know with 100 per cent certainty that if either of us was uncomfortable going into the ravine, then we wouldn't have entered. Our CRM training and our trust in each other meant that we both had to be happy before we entered — if I asked Harry to abort, I knew that he would.

Just before we entered the ravine, I called RCCNZ to update them. Once I was finished, Harry gave me clear instructions to stop communicating with the outside world for now so we could focus on the job at hand. It was just to be Harry and me — no one else mattered.

We had forward visibility of 50 to 100 metres as we made slow and steady progress into the ravine. We were literally flying from one

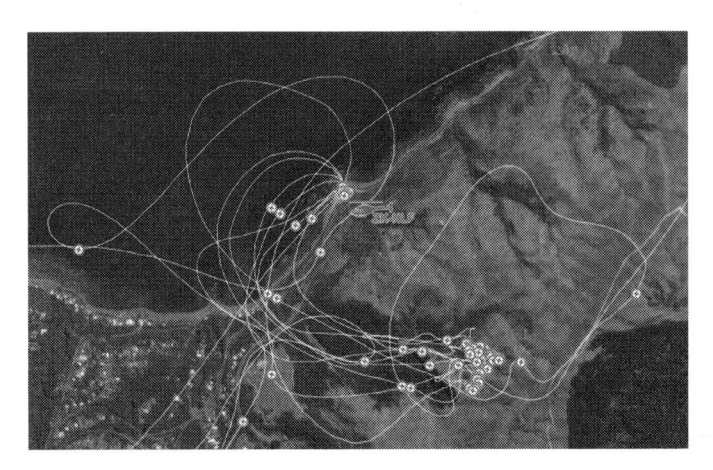

The Westpac Rescue Helicopter track during the Anzac Day search and rescue. *Life Flight*

stand of trees to another — get to a safe spot, identify another safe spot as far out as we could clearly see and fly to that. The tailwind was around 25 knots which I know made the flying more challenging for Harry.

I was not feeling any fear or worry as we made our way in. Although it was a long time since I had been in a situation where we were this low to the ground and surrounded by cloud, we had a solid escape plan, and were making good progress towards the strengthening beacon signal.

The cloud continued to lift and our forward visibility was slowly improving. We began to hear the second beacon locator signal, presumably from one of the crew's personal beacons. The signals were getting stronger as we moved forward — we didn't know what we would find when we got there, but we were getting closer.

I was elated when I suddenly spotted an airman making his way through the tall bush. At first I thought it must be a crew member from the missing helicopter, but as he was heading in the same direction as us, it made no sense. I pointed him out to Harry, who told me he heard from the other helicopter that a crewman had left his Iroquois and was running into the bush to try and find his mates. He was a rescuer, not a survivor. We discussed stopping and winching him up into our helicopter but decided to keep moving forward while the visibility was okay. We could always return and pick him up later.

Thirty seconds later we came to a slight fork — the DF unit was showing the signals coming from our right, so we headed up the right-hand fork. Seconds later we rounded a bend and spread directly in front of us across a large area was the wreckage of the Iroquois.

I felt sickened as I looked down at the crash scene. Parts of the crash site were still shrouded in the cloud and mist, making it even more eerie. This didn't look like a survivable crash. Helicopter debris was scattered for hundreds of metres, and it was clear that the helicopter had hit the hill at speed. We could see someone lying near the wreckage who was either unconscious or dead, but it was impossible to tell from the air. We could not spot anyone else from where we were.

As there was nowhere for us to land and we didn't have anyone we could winch to the scene, there was not a lot we could do from the air. Harry flew us back a few hundred metres until we were overhead the air force crewman. He was close enough to the crash site that it was quicker for us to direct him to the wreckage using hand signals than it would have been to winch him up.

We hovered above him until he reached the wreckage. He then waved us off, presumably so that he could look and listen for survivors without the noise and wind of our helicopter distracting him.

We headed back to the parking area where we had dropped off Lois and Lesley earlier to wait for word from him. The weather in the area had improved and flying out of the ravine was easier than when we had flown in.

Harry and I didn't say too much because there were no appropriate words. I called RCCNZ and the police to give them our update while Harry passed the bad news to the other Iroquois team via the aircraft radio.

As we approached the landing area, we saw it was now bustling with activity. The Iroquois was shut down in one corner so we landed in another. The emergency services were all gathering here, too, and the parking area had become the Incident Control Point.

It felt like hours had passed since we dropped off the medical team but when I checked my watch I realised it had only been around 30 minutes or so since they left in the police car. The air ambulance was probably not even airborne from Wellington Airport yet.

Before Harry had a chance to shut down the helicopter, one of the air force pilots ran over and advised us that their crewman had located two people — one alive and one dead — and needed help asap. The news that someone was alive was a welcome surprise and brightened up the bleak morning considerably.

I called ambulance comms to tell them a survivor had been found and we needed a winch-trained paramedic urgently. They advised that Pete Collins was a few minutes away. While we were waiting for him, Harry and I set up the helicopter for a winch rescue.

As soon as Pete arrived, we had a quick briefing then headed back into the ravine. By now the cloud had lifted significantly and we were able to fly directly to the scene. As we approached the crash site we spotted the crewman with the injured survivor, Stevin Creeggan, about 25 metres downhill from the main wreckage. The two of them were braced against a bush to prevent them from rolling down the hill. We winched Pete and his gear down to them and then moved off a bit so our downdraft and noise didn't affect them.

One of the light-hearted moments in an otherwise rugged morning was watching Pete's ambulance pack roll down the hill soon after he disconnected from the winch hook. As I watched the crewman race after it, I imagined Pete telling him: 'I need to look after him, you go get the bag.'

Wreckage of the Iroquois helicopter.

Pete did a quick assessment of Stevin, and although the mechanism of injury suggested that he would be more severely injured, all that was obvious were leg and chest injuries. He was conscious, responding to commands and had a good airway. Pete radioed us and we all agreed that we would use our nappy harness, not a stretcher, to winch him to the helicopter as we would need more rescuers on the ground to use the stretcher. Using the nappy would be quicker and allow Pete to continue searching for other survivors.

Normally, when we use the nappy harness, our paramedic would accompany the patient up to the helicopter. On any other day, someone as seriously injured as Stevin would be flown directly from the scene to Wellington Hospital, a 12-minute flight at most. However, there was still a chance of finding more survivors, and we were the only helicopter at the scene which could winch them out in a timely manner.

When they were ready, we moved into position and sent the winch hook down. When he was on the hook, Pete and the crewman helped guide Stevin out from under the bush and once he was clear we brought him up to the helicopter. I put a secondary safety strop on Stevin and left him sitting in the doorway for the short flight because it was obvious it would be too painful for him to be moved inside.

I didn't recognise Stevin as I sat next to him in the doorway, but given his bloodied and battered face and body, I most likely wouldn't have recognised him even if our paths had crossed before. Regardless of whether we had ever met, I felt a bond with him unlike I have felt on any other mission. Today I was winching him out of a helicopter crash; on another day, he could be winching me out. I wanted to give him a big hug but it looked like that would be too painful for him.

The plan was to load him onto an ambulance stretcher and send him on a 30-minute road ambulance journey under lights and siren. Unfortunately, I was unable to communicate my plan to the ambulance crew waiting in the parking area and we caught them by surprise.

They were expecting Pete to be with the patient, but it was only me and the patient sitting in the doorway. I signalled to the ambulance officers to come over to us but they didn't move. I found this very frustrating, even though, in hindsight, they were right and I was wrong. Ambulance officers are trained to never approach a running helicopter unless they are escorted by someone from the helicopter. They did exactly what they were trained to do, and no matter how much I waved, they weren't approaching.

Instead of taking the time to go over and explain to them what was going on, I waved over one of the Iroquois crewman and we took the ambulance stretcher from the startled ambulance officers. We brought it to the helicopter, and then as gently as possible, used the winch to lower Stevin onto the stretcher.

We rolled the ambulance stretcher back to the ambulance officers and I gave them the best patient handover I could. I told them the helicopter needed to remain on scene and they needed to drive him through to the hospital.

They did an absolutely sterling job, particularly considering that neither of them was an intensive-care paramedic and they had a seriously injured patient thrust at them unexpectedly. They headed off to hospital, asking for a more highly trained paramedic to meet them en route.

Once they headed off, Harry and I waited for word from the crash site. A few minutes later, word came. There were no other survivors.

THIRTY-FOUR
ANZAC DAY — PART 3: THE AFTERMATH

THE NEWS THAT there were no other survivors quickly spread across the landing area and shattered the hope that had built after Stevin was rescued. Everyone seemed to stop what they were doing to take in the news. It was a poignant moment of silence on the very day we remembered those who had sacrificed everything for New Zealand.

As news of the helicopter crash got out, my cell phone and pager went nuts. Just about every news agency was texting or calling me, but I ignored them all. A military helicopter crash on Anzac Day was big news, and everyone wanted to know what was going on, but this was not the time to share any information.

Soon after the police inspector in charge asked us to winch one of their officers to guard the crash site as it was now considered a potential crime scene. The police also wanted aerial photos of the crash scene and the surrounding area so they could make plans for accessing the area.

We winched Pete back up to the helicopter after we set the police officer down. The air force crewman elected to remain at the crash scene with his dead mates, and I cannot even begin to imagine how he was feeling.

When we returned to the landing area, I finally had a chance to talk to some of the Iroquois crew that had been searching alongside us. Out of respect for the guys on board and their families, they wouldn't tell us the names of the three deceased, but they confirmed that it was not one of the pilots or crew that I knew well.

Over the next hour or so we carried out a few different tasks for the police and air force. Another Iroquois arrived from Ohakea with an accident investigation and SAR team. I had met the head of the investigation team before and we discussed what involvement they might want from us for the rest of the day.

He felt confident that the team he had flown in with were up for any tasks such as winching in investigators, but he asked if we could return later when it was time to winch the deceased aircrew from the scene. I spoke with Harry, who was happy to come back if the weather was okay.

WE RETURNED TO our base to clean up, refuel and be ready for our next call-out. Colin had come into base to answer the emergency hotline and assist us in any way he could. I called the hospital and was thankful to hear that Stevin had made it alive. He was currently in ICU and due to head to theatre shortly.

Our decision not to fly him directly to the hospital seemed okay, at this point anyway. My logical brain told me that in such a high-speed accident where three people had died and the sole survivor had so many serious injuries, he was likely to die, too. Once we got him out of the ravine, anything else we did, or didn't do, would probably not affect his outcome. Unfortunately for me, this was not a logical problem — it was an emotional one.

Harry and I sat down and had the first of many discussions about the mission, and we were confident that we had handled the morning well. There had been a lot going on, many decisions to make and a lot of teamwork required. We agreed our CRM was extremely good and that we worked well as a team. Neither of us ever felt unsafe or that we were pushing our own safety boundaries too far. It had been a terrible and distressing day, but we felt that we performed our parts well.

Media interest had been intense from the moment it was known an air force helicopter had crashed but we weren't making any media comments until we were given the okay from police, RCCNZ and the air force to talk about our part in the rescue. I spoke with my CEO and he was happy for me to manage the media in my normal way.

Before doing any interviews, Harry and I discussed what we would say. We agreed we would avoid answering any questions about weather or other conditions that could be used to attribute blame or speculate on the cause of the accident. We would stick to the facts of the things we did and the challenges we faced. As we were not there when the accident happened, any speculation on our part was irrelevant.

TV3 called and asked if they could send out a team to interview us and pick up our video footage. Due to their sponsorship agreement, they were the only ones entitled to use our video from the crash scene — if we released it. We had some dramatic winch camera footage, but there was no way in hell I was going to release that on the day. I felt everyone needed at least one night to put things in perspective before we could even begin to ask if it was okay to release it. I told them that an interview was okay, but that nothing would be released until we had clearance from the air force and family members, and that could not happen until the next day at the earliest.

When the fixed-wing air ambulance returned from Blenheim later that morning, I spoke with the mother of the teenage ICU patient. She totally understood why the fixed-wing was used instead of the helicopter. In the end, her son had to return to the operating theatre in Wairau for emergency surgery before he could be transferred. This had delayed things by quite some time. It was a huge relief to know that our decision to send the fixed-wing air ambulance didn't further delay his transfer.

Harry was in the middle of a TV3 news interview when I got a shocking phone call from Gordon McBride, the TV3 Wellington bureau chief. Gordon told me that it had just been announced that Stevin had died in hospital. I was astounded and told Gordon that

I couldn't believe it as I had spoken to the hospital only minutes before the news crew arrived and they told me that he was alive. I interrupted Harry's interview to give him the news — it was particularly relevant since he was speaking about how glad we were that there was one survivor.

Internally, I started to review the decisions we had made that morning and wondered if any of these had adversely affected Stevin's outcome. Did using the nappy instead of a stretcher make things worse? Sending him by road ambulance was the hardest choice of all. Would he still be alive if we had flown him directly to hospital?

I called back ICU and told them that it had just been announced Stevin had died. The nurse I was speaking to, who was a friend as well as one of the flight nurses, assured me that he was very much alive, in fact she was looking at him as we spoke. Harry and I were both greatly relieved to hear this, and he continued his interview as though Stevin was still alive. At the very end of the interview, they filmed a couple of questions about the mistaken statement, just in case.

After TV3 left, Harry and I continued to pick apart every detail of the operation and in the end we came to the same conclusion. We stood by the decisions that we made and we would have no problems defending these decisions in a Coroner's Court or any type of inquiry into our actions.

Thankfully, we had no other emergency call-outs during that day, but my phone never stopped ringing. The media calls were relentless, and throughout the day Harry and I spent a lot of time doing interviews.

Low cloud prevented us from returning to the crash site to winch out the three who didn't survive the crash. I didn't personally know the pilot, Hayden Peter Madsen, co-pilot Daniel Stephen Gregory, or crewman Benjamin Andrew Carson, but their deaths have left a permanent scar on my soul.

Even though Harry and I didn't end up winching them from the scene, I still feel humbled and honoured that the air force asked us to do it. It was a long and emotional day, and by the end of it I was ready for a decent meal, some good wine and a hug. My close friends ensured I had plenty of all three.

THE FOLLOWING MORNING, Gordon from TV3 called again wanting the video footage of the rescue. Harry and I had reviewed it and he was happy, from an aviation point of view, to release it. I told Gordon we had to get permission from the New Zealand Defence Force, the police and Stevin's family before I would release it. I contacted the air force, and they said that if I got consent from Stevin's partner, who they said was his designated guardian, then they were okay with us releasing it. The police said the same.

I then headed to the ICU unit to get the needed permission. His partner was sitting with several of Stevin's air force mates and I showed them the footage. They obviously found it upsetting but were grateful that we had been able to rescue him, and as they knew there was a lot of public interest in the crash, agreed for it to be released.

Even though I didn't technically have to, I also showed it to Stevin's parents, Gaile and John, who were in the waiting room. I didn't need to understand the family dynamics because whether they were his legal guardians or not was immaterial to me. They were his parents, sitting in the waiting room, unsure if their son would live and they deserved to see the footage before it was broadcast on TV. This turned out to be a very good decision on my part, because over the next few days it became known that Stevin's parents were actually his legal guardians, not his partner!

Once I had the consents I needed, I told Gordon we would release the footage but asked to have a chance to explain our decision-making during the news story. The three things we knew could be questioned, particularly by the armchair critics, were the use of the nappy harness instead of a stretcher, sending Stevin up the wire without the paramedic, and electing to use a road ambulance to get him to hospital instead of flying him there directly.

Even though I didn't know Stevin prior to the accident, I had formed a real emotional connection to him and the accident. I began to visit Stevin and his parents in hospital every day and I have formed a very special friendship with the three of them.

One person who went up in my estimation was John Key, the Prime Minister at the time. I was in the hospital and popped into ICU to visit Stevin. As I approached his room, I saw that the PM was there visiting Stevin and his parents without any reporters, TV cameras or photographers. It was not a visit for the media, it was simply a visit from a fellow parent. It made me very proud to be a Kiwi.

HARRY AND I were invited to the combined funeral for the three aircrew at Ohakea Base later in the week.

Three Iroquois arrived, each carrying the body of one of the aircrew. The sound of them arriving is forever fixed in my mind, as is the sound of the bagpipes which led the procession of caskets into the large hangar where the service was held. To say it was an emotional service would be an understatement.

The funeral was followed by an immense celebration of the three men's lives. It had been a big week full of highs and lows. The week had such a sad outcome for so many people, but there was a survivor and we did our best for him.

Stevin was in the ICU for more than a week before he was released to a ward. Eleven days after his accident he was doing well enough to be transported to Palmerston North Hospital, the hospital nearest to his home.

We would be flying him on the fixed-wing, and I arranged to be on the flight with him and his mum. He was very excited to hear that he would be flying, especially because he thought it would be on the helicopter. I was surprised that he wanted to get back on a helicopter so soon after the crash but I told him I would see what I could do.

I asked if he'd be okay to have the media cover the return flight and he was keen to do anything he could to support Life Flight.

I spoke to my CEO and he was happy for us to use the helicopter, especially as I said that I could get some positive media coverage out of it. Stevin was really excited about the helicopter flight, as was his mum when I told her there was room for her, too.

About an hour before the flight, I had a chat with Mike, our duty pilot, and asked him if he could stop in at Ohakea on our way to Palmerston North. It was only a bit out of our way, and I thought it would be a great morale boost for Stevin if he got to see some of his mates.

Mike said yes, so I asked Stevin, who thought it was a great idea. An hour or so later we headed off on the 40-minute flight. I truly admired Stevin for his willingness to get on a helicopter so soon after being involved in a major crash, and he handled it well. His mum was up front with Mike and was enjoying her flight a lot. The flight was uneventful, the way it should be, and there was an air of excitement as we approached Ohakea for our quick visit.

We landed in our designated spot and within minutes an enormous group of people came to greet us. Stevin was unable to get off the stretcher, but we pulled the stretcher out of the helicopter so that people could come up and say hello.

It seemed that just about everyone on base came out to say hi. I was told that seeing Stevin alive and on the road to recovery was a great morale boost for the entire base. After dealing with so much grief and sadness, they were pleased they could see and speak with him. After being on the ground for about 30 minutes, we loaded Stevin back into the helicopter for the short flight to Palmerston North.

THE ANZAC DAY rescue resulted in Harry, Pete and me receiving a 'Safety in the City Award' from the Wellington City Council. The three of us, plus the air force crewman who left his helicopter and ran through the scrub and bush on Anzac Day, also received a New Zealand Search and Rescue Award.

Although both awards were nice, nothing was as satisfying as watching Stevin slowly and purposefully getting on with his life. Even more special is the way in which I am greeted by his parents each time I see them. The way they look at me and the hugs they give me are the best reminders of why I so loved the job I did on the helicopter. The difference our teams make to other people's lives is indescribable.

It still seems amazing to me that Stevin survived the crash and was eventually able to walk out of hospital, albeit with a Zimmer frame. His list of injuries is unbelievable: 18 fractures to face; tear drop

Dave, Harry Stevenson, Stevin Creeggan and Pete Collins at the safety award ceremony.
Dominion Post

fracture to C2 vertebra; flail chest; haemo-pneumothorax; collapsed left lung; lacerated spleen; fractured neck of femur; fractured mid-shaft of femur; T11, T12 and L1 fused due to crushed/compressed and fractured vertebrae; traumatic dislocation of right elbow; fractured sternum; fractured left shoulder blade; fractured bones in right hand; fractured bones in right foot; burns to both hands and face; lacerations to face; penetrating injury to left eye; soft tissue damage to right shoulder; soft tissue damage to right knee; two fractured teeth; and traumatic brain injury.

Stevin will always be a mate as well as a reminder of why I was so lucky to do the job I loved for 25 years.

THIRTY-FIVE
SUICIDE AND PTSD

IN NEW ZEALAND, every year about 500 people will take their own lives and for most of these, someone in the emergency services will attend the aftermath.

It is the darker, hidden side of our job and one of the most harrowing. My life has always been about rescuing people and I've always found it difficult responding to a mission when someone has chosen to take their own life.

The saddest thing about suicide is that the person who takes their life does it because they think it will remove their pain and make life easier for others. In my opinion it doesn't. It simply passes the pain on to someone else.

Going to a scene where someone has taken their own life has never bothered me on a physical level; a gunshot wound is a gunshot wound no matter who pulls the trigger. However, they disturb me on an emotional level because I've never understood what would drive someone to such despair.

Perhaps it is simply because I have never been in such a dark place that I ever contemplated taking my own life. I count my blessings and thank the people around me — my co-workers, friends and family. They are constant reminders of how much I have to live for, even when I have gone through rough patches in my life.

There are two suicides, in particular, I've never forgotten.

One was a farmer found in his shed on a beautiful coastal property on a sunny spring morning. He was living in a stunning place surrounded by all sorts of boy toys: a boat, a jet ski, scuba-diving equipment, fishing and hunting gear. From the outside, he seemed to have it all. He is a reminder for me to never judge how good or bad I think someone else is doing because we are all battling demons no one else knows about.

The second was a man who swallowed a large quantity of poison, expecting a quick and painless death. We were called to transfer him from a regional hospital back to Wellington ICU. He was conscious and alert when we arrived and I had a chance to talk with him and get to know him a bit. He told me how bad his life was, how nobody loved him and how there was nothing to live for. Over the next couple of days, his family and friends from all over New Zealand and Australia gathered at his side. Finally, he realised how loved he was and decided that he didn't want to die. Unfortunately, there was no way to reverse the effects of the poison, and he died a slow and painful death, surrounded by people who cared.

He reminds me there are some people who believe they are alone in this world, but there is always someone who cares; sometimes you just need to seek them out.

LIKE ALL MY fellow emergency workers who are out there every day, I have seen some horrible things over time. People mangled amongst twisted metal or burned beyond recognition. Families brought to their knees in distress as they watched their relative pronounced dead. Like most of my colleagues, I have always found jobs involving infants and young children the hardest. I will never forget pleading with a mother to let go of her baby so I could perform CPR, only to have the baby pronounced dead after we arrived at the hospital. It was one of the hardest missions I have ever been involved in and one I prefer to keep buried in that little space in my head that I don't visit often.

Today, my social media feeds are full of stories about police officers, firefighters and ambulance officers, from across the globe, suffering from Post Traumatic Stress Disorder, or PTSD. Sadly, many of these wonderful, caring people end up taking their own life without getting the help they desperately need.

PTSD affects men and women, young and old, newcomers to the field and people with more than 30 years' experience. Working in the emergency services, we all know we need to find a useful and non-destructive way to cope with what we see on the job, because we all have the potential to find ourselves in such a dark spot that suicide seems the only option.

I was first exposed to PTSD before it was even labelled as a disorder. Back at West Hamilton Beach, one of the firefighters at my firehouse was a Vietnam veteran. He didn't talk about it much, but he was clearly struggling with some of the things that he had seen or had done while over there.

A lot of the news coverage at the time reported veterans turning to drugs or alcohol, suffering from nightmares, committing suicide, or committing domestic violence. It was frustrating to know that he was suffering, but no one, especially teenage me, had any idea how to help him. Most of the older guys just muttered unhelpful things like 'grow some balls' or 'just get over it'.

'Get over it' or 'deal with it' was the way I was taught to deal with things. It's only in the last 10 or 15 years that people have been more open with their feelings and encouraged to talk with a friend, or to get professional help, if they were feeling distraught about an incident.

I often wonder which of my friends, whether they are in the emergency services or not, might be silently suffering after a difficult job or an accumulation of some hard days. I hope that all of you know that my phone is always on, and I am always available to talk if you feel the need to.

I would like to think that we all have someone to turn to, sometimes all you need to do is be willing to ask.

I FEEL LIKE I am one of the lucky ones because I have always found ways to cope with what I have seen and dealt with.

Following our helicopter crash in 2003, everyone on board the helicopter, as well as everyone managing the aftermath, was required to go for a mandatory counselling session. I went to see a police counsellor named Sue. We chatted comfortably and after about 15 minutes she seemed satisfied that I was okay with the aftermath of the crash; everyone had survived, we had a temporary helicopter and a replacement helicopter would arrive by the end of the year.

As I had a one-hour session scheduled, she offered to keep on chatting. She asked me how I coped on a day-to-day basis. My initial response was something along the lines of 'I just do'.

She then explained that everyone has a coping mechanism. Some people drink, some do drugs and some, unfortunately, beat their partners, but we all have a 'go to' method of coping. She really wanted to know what mine was.

I thought about it for a few minutes and then started laughing. My friends, the gym and 'chick flicks' was my answer.

Ever since I was a teenager, around the same time I started at the firehouse, I have often cried like a baby at sad scenes in a movie, or even sad songs. It became rather embarrassing so I would warn people I was going to the movies with for the first time, that they should just ignore me if I cry.

I didn't think it was a great look for a supposed big brave rescue helicopter crewman, but it was my way of releasing the stress and grief that built up on a regular basis.

My closest friends have no idea how many times I have shown up at their house, presumably to bum a free meal, but really just to feel the love and warmth of people that care about me. Very rarely did I have to talk about my day; for me the release was just being with them, often not saying a word. A hug, a glass of wine and the chance to hang out with them and my life was good again.

You guys all know who you are, but you will never, ever realise just how big a difference you have made in my life.

THIRTY-SIX
THE SPECIAL PEOPLE IN MY LIFE

LIKE EVERYONE ELSE, I have had highs and lows throughout my life. I know these are part of the ebb and flow of life and looking back they have helped make me the man I am today.

A few days before Valentine's Day in 1994, I married Jenny, a special woman I had met through Money and You. Two years later, almost to the day, we lost a baby midway through her pregnancy. Losing the baby, that far along, will always rate as the worst period of my life. Unfortunately, our marriage didn't survive this tragedy, but luckily, our friendship did. Today, Jenny and her husband Paul are two of my closest friends.

Bessie helping out with a Life Flight street appeal.

A couple of years later I became close friends with an ICU nurse named Clare, who I eventually convinced to date me. Ironically, she was a Kiwi with a desire to live in America, and I was an American with a desire to stay in New Zealand, so our relationship ended when we became separated by the Pacific Ocean. I still count her as one of my close friends, and love catching up with her and her family when they visit.

Besides those two special people, the only other female I have loved since I moved to New Zealand was my dog Bessie. She was the best dog ever! My ex-wife and I shared custody of Bessie, and that is probably why we are still friends today. If we didn't speak to each other regularly to get the dog moved between our houses, we probably wouldn't have had a reason to stay in touch. One of the saddest days of my life was the day Bessie became so ill that putting her to sleep was the kindest thing to do. It broke my heart, but I laid beside her at the vet as she took her last breath.

Today I have almost 500 Facebook friends, but I also have a smaller, close-knit group of people I choose to spend most of my free time with.

If they hadn't banned me from doing so, I could write an entire chapter, if not a book, about my closest friends and what they mean to me. These friends are spread across the world, from Wellington to Seattle to Dallas to New York. Of course, I include my relatives — my sisters and their husbands in this group — but the others are also family. They are the people I have chosen to include in my life, and they have allowed me into theirs.

These friends have helped me celebrate my highs and got me through the lows. They are the ones who have supported me during my worst days at work just by being there. When Dad died, they were the people who gathered at my side and even offered to fly to New York with me. On the rare occasions when we have got offside with each other, we have worked hard to make things right because our friendship was more important than everything else.

I know that you all know who you are so I will simply say thank you for always having my back, and know that I will always have yours!

Dave and his dad, Bill, in 2014.

Dave and his mum, Judy, in 2015.

Doug, Dave, Trish, Tara and Declan at Disneyland.

Georgia, Bettina, Grant and Zara.

ONE OF THE sad things about moving from New York is that I have never lived in the same city as any of my nieces or nephews. I have always been grateful that my sisters worked hard to make sure that their kids, Jason, Tracey, Jaclyn, Michael and Amanda, knew who I was, even though I only saw them occasionally and was rubbish at staying in touch in between.

In New Zealand, I have had the privilege of being 'Uncle Dave' to four fantastic kids, the children of my closest friends.

Doug and Trish are the proud parents of Tara and Declan while Bettina and Grant are the equally proud parents of Georgia and Zara.

The best gift that the kids, and their parents, gave me was showing me that there was life outside of Life Flight. Work had always been my highest priority, and I would run out of any dinner or party if the pager went off and I was needed for a helicopter call-out.

The first time I babysat Tara that all changed. On that night, I realised that no matter what happened, no matter how much I might be needed for a helicopter call-out, I couldn't go; I only had one priority, and that was babysitting this precious little girl. It was an eye-opening experience, and I loved spending time at their house, being part of their family that grew to include Declan a few years later.

By the time Georgia and Zara were born, I had managed to reduce the number of hours I spent at work, a bit, so I got to spend a lot of time with them as they grew up, through all the stages of their young lives, and I have forged a strong bond with them.

Seeing life through a child's eyes has been amazing, especially as their view of the world changes as they grow older. I have spent countless hours at zoos, parks, school shows and cafés with these kids and I have loved every minute of it.

The favourite thing I have created with Georgia and Zara is adventure days. Instead of buying them each a present for their birthday or Christmas, I gift them an adventure day: a day where I go out with them individually and they get to choose everything that we do from the time we leave the house until we return home after dinner. I never knew how many activities a kid could pack into a day! These days are exhausting for us all, but they are my four favourite days of the year where I get to hang out with my favourite little people.

These two families are the closest of my close friends. I have had the fun of travelling with both families to the USA, at different times, and was even trusted to accompany Tara and Declan home from San Francisco, so their folks could have a few days' extra holiday. I have become quite a good 'manny' for Georgia and Zara, babysitting for several days in a row to give their folks time for a break.

Not having had my own children, it has been very special being part of these children's lives. What the kids, and their parents, have given me is the gift of family, and being part of the village that helps raise a child.

THIRTY-SEVEN
CHRISTCHURCH EARTHQUAKE

MOTHER NATURE HAS provided us with many missions over the years — including southerly gales, cyclones, flooding and landslips. As a smallish island in the middle of the vast Pacific Ocean, New Zealand gets its fair share of interesting weather.

But on 22 February 2011, Mother Nature threw a whole new problem our way. I was sitting having lunch at the base when I noticed the helicopter's blades bouncing as a minor earthquake shook the building. Wellington is subject to frequent tremors, so this was nothing unusual and I didn't bother even getting under the table.

It was, however, a large, shallow earthquake near Christchurch, 300 kilometres to the south. Within seconds my Twitter and Facebook feeds were flooded with alerts of a devastating earthquake that had caused massive damage. Mike, the duty pilot that day, had family there and was worried for their well-being.

Within 40 minutes of the quake, the head of the Christchurch St John ambulance service called. By chance, he was in Wellington along with the Christchurch heads of the Fire Service and Police, who were, ironically, speaking at a conference about their response to the earthquakes which had struck Christchurch the previous September. They needed us to fly them to Christchurch immediately.

In case we couldn't refuel in Christchurch, we loaded the 300-litre long-range fuel tank in the helicopter, and as soon as the service heads arrived, we set off. As we crossed Cook Strait, the guys in the back started to get texts and emails alerting them to how serious the situation was. Buildings were destroyed, vehicles crushed and there were many more deaths than the media was aware of.

As we neared Christchurch we saw smoke rising in the distance. Flying above the northern suburb of Burwood, we could see thick grey mud flowing through the streets, caused by liquefaction. The CBD was a scene of destruction and ruin. City streets I had walked through just a few weeks before were packed with rubble and scrambling people. A tall building had collapsed and was on fire, surrounded by emergency service vehicles — this was the CTV building where many people died.

After a quick aerial tour of the city, we dropped the three service heads off and landed at Garden City Helicopters, the home of the Christchurch Westpac Rescue Helicopter. After refuelling they told us to head to Hagley Park in the centre of the city, where all the rescue helicopters were gathering to await further instructions. The ambulance service had set up triage tents in the park, trying to reduce the number of injured being sent to the nearby hospital, but, surprisingly, they were not busy. Most of the injured were already in hospital or at other triage points.

Our view of the CTV building.

Nine rescue helicopters from all over the country ended up sitting in the park. Most of the casualties were in the city area so rescues in the outlying suburbs were not needed. The plan to fly patients out of Christchurch, to free up beds in the hospital, was thwarted when low cloud closed in over the city.

At dusk, along with several of the rescue helicopters, we relocated to Garden City Helicopters at the airport. I recognised many people there, including pilots and crew that I have flown with in the past. Somehow Simon Duncan, Garden City's manager, was able to wrangle up a huge feed of fish and chips from a nearby shop that hadn't lost its electricity. Almost everyone at the hangar lived or had relatives who lived in Christchurch and they were all concerned for their well-being.

Most of the others headed off to sleep at a friend's or relative's house, but I remained at the airport to help any fixed-wing air ambulance crews that might arrive once the runway was reopened. It was a long, sometimes terrifying night, as large aftershocks struck every few minutes. I was worried about the USAR (Urban Search and Rescue) teams which were crawling through partially destroyed buildings looking for survivors; I knew many of the team members and could not imagine what they were feeling.

Gail, Garden City's flight coordinator also stayed overnight, and every time I woke with a fright from the latest aftershock, she would say something like, 'Nah, that wasn't very big.' I couldn't imagine what the city had experienced if these were considered small! I think there were only two times that she said, 'Yeah, that was a biggie.'

Starting overnight, and continuing for days afterwards, fixed-wing air ambulances from all over the country arrived to transfer patients out of Christchurch. Early the next morning our helicopter wasn't required for any flights so I took an opportunity to jump into an ambulance support vehicle and drive into town with Tony, one of the St John paramedics. Being among the devastation was different from seeing it on TV or even from the helicopter. The hardest stop of the morning was at the CTV building; as bad as it had looked from the air, it was worse close up. Many firefighters, police and ambulance officers who had been there since the earthquake struck the day before were desperately trying to rescue those still trapped in the building. It was distressing to witness. In the end, 115 people lost their lives in the CTV building.

THAT AFTERNOON, WE were tasked to transfer a seriously injured woman from Christchurch ICU to Wellington ICU using a Wellington Hospital medical team who had arrived on a plane earlier. I knew a lot of the doctors and nurses in Christchurch ICU and they were so grateful for the support they were getting from around the country. I wished there was more we could do; they were all here helping strangers while their own homes lay in ruin.

Normally, we would land in Hagley Park for hospital transfers, but as there were no road ambulances available we landed the helicopter on the street outside the hospital, loaded our patient and headed home. I was glad to be getting back to some normal turbulence and not shaking ground!

I was tired, hungry and in need of some company by the time we got back to Wellington. My friends, as always, were there to support me, and I had dinner with Bettina, Grant and the girls on Lyall Bay beach. It seemed so surreal to be enjoying such a normal dinner on a lovely evening knowing the total chaos that was going on just 300 kilometres away.

The February Christchurch earthquake was the worst in terms of destruction that New Zealand had experienced for a long time. I was disappointed that we couldn't do more to support our southern friends on the day, but I was extremely proud to be part of the network of air ambulances that helped so many people in the aftermath.

THE CANTERBURY EARTHQUAKES were the first of many earthquake sequences that the country has experienced since 2010/11. Subsequent earthquakes in Seddon, also on the South Island, were closer to home and did some damage in Wellington.

I spend a lot of time speaking and working with families and organisations about how to be best prepared for the next earthquake, because there is certain to be one. New Zealand really is the shaky isles.

On 14 November 2016 a 7.8 magnitude earthquake, one of the strongest to ever strike New Zealand, ripped through Kaikoura, an area midway between Blenheim and Christchurch on the South Island.

This earthquake caused major damage to the tourist area of Kaikoura as well as to buildings in Wellington. Several buildings in Wellington were damaged beyond repair and have since been torn down.

A 7.8 magnitude earthquake centred under Wellington would do significantly more damage to the region and its infrastructure than the Kaikoura earthquake did. My hope is that people take notice of these other earthquakes and prepare themselves, their families and their organisations for whatever Mother Nature throws our way.

THIRTY-EIGHT
POLICE MISSIONS

FOR ME, THERE is little that will ever compare to the feeling of accomplishment and satisfaction that comes from being part of a life-saving team. Looking back on my career I am immensely proud of what I have been a part of, both on the road and in a helicopter.

That being said, some of my best-loved times in the air have been when our helicopter was used to assist the police in crime-fighting activities, or on the few occasions I had the opportunity to ride along with a police helicopter.

There is only one dedicated Police Aviation Unit in New Zealand, which is based in Auckland. If the police in other regions require air support then a local helicopter is used. In the Greater Wellington region, our helicopter was often called upon to assist.

We were also used to transport the Wellington-based New Zealand Army Bomb Disposal team, including a bomb disposal robot, to the scene of suspicious packages around the middle of the country, anywhere from New Plymouth or Hawke's Bay down to Marlborough or Nelson.

Over my career, I have been involved in many different types of police missions — car chases, aerial surveillance, searches for escaped prisoners, and deploying specialist squads including the Armed Offenders Squad (AOS) or Special Tactics Group (STG) — our equivalent of SWAT teams. I have also been very fortunate to have had the opportunity to ride along with police helicopters in Auckland, Los Angeles and Nassau County (New York). My night shift over LA was an eye-opening, fantastic experience.

While I can always count on a chick-flick to bring on a few tears, my movie genre of choice has always been action and adventure. Any movie with a good helicopter chase is enough to keep me interested for at least a couple of hours. And as much as I enjoy a good chase scene in a movie, being part of the real thing was even better.

ONE OF MY favourite nights in a helicopter, ever, was with the Los Angeles Police Department (LAPD) in 2003. We were looking at purchasing a used Forward Looking Infrared (FLIR) system for our helicopter, and I thought it would be useful to be able to see a police helicopter using a system.

Like NVG and GPS, FLIR is another example of a technology that was designed for the military and then adopted for use in the civilian world. A FLIR system uses a thermal imaging, or heat detecting, camera system to display a picture of different levels of heat sources on a monitor. If you have ever seen a news story or a scene in a movie where a video display shows a fuzzy image of a person hiding behind a bush or under a car, most probably a FLIR unit was being used.

When I flew with them back then, the LAPD had, and I assume still has, a very active aviation unit. At the time, there were anywhere between two and four helicopters flying routine patrols over Los Angeles 24 hours a day. On top of that they had many other helicopters for specialist work.

I was assigned to one of the four patrol helicopters, a Squirrel AS350B. The two pilots were also police officers, and between them flew the helicopter and worked the numerous radios, their moving map display and their FLIR unit.

I jokingly asked them if I should be worried that they were in bullet-proof vests and I wasn't; they replied that they very rarely got low enough to be shot at, so I should be all right. Now I understood the waiver I had signed before I got on board!

Police Missions **229**

The shift started in daylight, around 4pm, and I got a real feel for how massive and spread out Los Angeles is. Down below, the streets and highways were all congested with traffic. The city was covered with a layer of smog which was very noticeable from the air.

Over the first 30 minutes I noticed that they were flying very 'gently'. When I mentioned this they said that they always flew less vigorously when they had guests on board so they didn't need to clean up vomit. I told them that they didn't need to worry about me; whatever they did, I would be fine. One of the guys laughed out loud and said that was what all their guests said, but that they had managed to make everyone air sick, from city councillors to their toughest SWAT team members, and they weren't going to take any chances.

A few minutes later we were called to be air support over a house with a burglar inside. Our job was to circle the house, over and over and over again, in case the burglar made a run for it.

As we circled I quickly got terribly bored and picked up the stabiliser binoculars that were sitting on the seat next to me and started having a look around. One of the cops asked the other to get a 'tag number' (licence plate number) off a pickup truck near the house, and I piped up 'got it' and told them the number.

They both turned around and once they saw me on the binoculars they realised that I really wasn't going to get sick. They then went from worrying about me getting air sick to showing off, doing all sorts of weird and wonderful things with their helicopter. The next six hours I had the best helicopter time of my life, flying low level around Los Angeles watching them use FLIR, moving maps and their Nitesun for different jobs.

At one point, we headed to an armed robbery where there were two people who had done a runner. We got overhead as one was apprehended, and quickly located the other with the FLIR unit, hiding behind a dumpster. After they talked to the ground units and led them to the suspect, they got a radio call to say the suspect no longer had a gun. In a very impressive and surprising display the guys started searching roof tops between the scene of the robbery and the spot where the suspect was caught.

When I asked them why, they said that people quite often tossed their guns onto a roof, but there was enough heat left in the gun handle to be detected by FLIR. Sure enough, after a short search, they detected a slight flicker of heat on a roof, and it turned out to be a gun.

During the shift I learned that a lot of the benefits from FLIR came from the experience of the operator as well as the technology itself. I proved this myself when they gave me a go at operating the unit when we were trying to find a mugger in a park. At one point I zoomed in on a heat source, obviously a man out walking his dog. I pointed this out to them and they chuckled and told me they had a name for those people — they were call K9 units. It was not a man walking a dog, it was a police dog and his handler out searching for the suspect! It was nice I could keep them amused.

This was a night to remember, and a special treat that I will always treasure.

LIFE FLIGHT ENDED up purchasing a pre-owned FLIR system which we used for several years to assist with search and rescue. On a dedicated police helicopter, the FLIR system would always be attached, but we only attached the system to our helicopter when it was required. It would take us about 20 minutes to attach the external camera and internal display and computer unit, but we used it on several SAR missions with good results.

During one operation, the police were looking for a person who had threatened to commit suicide. We were called in with the FLIR unit and were able to spot them among some thick grass. We could then communicate with the police and talk them over to the location. Without FLIR we never would have spotted the person.

FLIR technology continually evolves and unfortunately, like most computer systems, older units like ours quickly become obsolete and unable to be maintained. When our helicopter-mounted system reached that point we moved to a hand-held FLIR unit, which was much less expensive and continued to be a useful SAR tool.

✈

MANY OF THE police missions we helped with involved armed police and police dogs. We had a dog harness which we could use to winch a dog in or out of the helicopter safely, alongside its handler.

Over my years in the job I had a few injuries, mostly just cuts, bruises and the occasional sore back, but after 25 years, my most serious injury ended up being a dog bite.

The bite occurred when we were deploying a drug team to raid a house in a remote area. We had four armed police officers, one of whom was a dog-handler, and a police dog, a Labrador retriever, on board. As we approached our landing zone we spotted an unleashed pit bull in the area. Normally a dog runs from the sound and wind of

The Westpac Rescue Helicopter performing a display with the police Armed Offenders Squad at a Police College Open Day. *Bruce Pool*

a helicopter, but not this one — as we came in to land the dog came towards the helicopter, barking and showing some rather large teeth.

The police did not want to abort the landing because their surprise raid was no longer a surprise, so we landed. As the skids touched the ground I jumped out, keeping an eye on the dog as I opened the rear cabin door to let the police out. As soon as the police dog got out of the helicopter the pit bull attacked.

The two dogs got into an aggressive fight and instead of doing the smart thing, letting them sort their stuff out, I stupidly decided to help the dog-handler try and separate them. By the time we finally managed to separate the dogs and tie up the pit bull, both the handler and I ended up being bitten. The police dog was okay, and neither the handler nor I was seriously injured, so they decided to proceed.

When I was back in the helicopter I found that I had been bitten on the arm, hard enough to draw a good deal of blood. Nothing too serious, but I knew I would have to get myself into hospital to have

A police dog and heavily armed AOS team member in the back of the helicopter.

it looked at after we got back to the airport. But before we got back to base we were called out to the first of several road traffic crashes that afternoon. It was during my third trip to the Emergency Department that day that I finally had a chance to have the bite looked at by a doctor. The most painful part of my day ended up being a tetanus shot!

NZDF bomb robot coming out of the helicopter during a real deployment.

NZDF robot being put into the helicopter during a training exercise.

BRINGING THE NEW Zealand Army Bomb Disposal Unit and their robot to the scene of possible explosives did not happen often, but it became an important part of what we did.

The army developed a set of ramps that was used to manoeuvre the robot into the back of the helicopter through our rear clam-shell doors. Unlike many of our missions, most of the helicopter training for them could be done on the ground because the difficult task was getting the robot in and out of the helicopter.

I took part in about half a dozen real deployments over the years. We would fly the team and their robot to the scene, and then as soon as they were dropped off we would return to Wellington empty. The urgency was getting the team to the location — once they were finished they would be driven back to Wellington in the back of a van.

In an ironic twist, one of the army guys I flew with a couple of times absolutely hated flying in a helicopter because he didn't think they were safe. He was happy as could be to work with explosives, but helicopters scared the hell out of him. To each his own!

FOR MANY YEARS, we trained with the police Special Tactics Group (STG) on an insertion technique known as 'fast-roping'. When fast-roping, a helicopter flies above the area that a team needs to get into quickly, and a thick rope, attached to the helicopter at one end, is dropped to the ground. The STG team members then slide down the rope, like a fireman's pole.

Fast-roping is much quicker than winching and is an ideal way for the police to get a team of people into an area rapidly, when a helicopter can't land. I was involved in many STG fast-rope training flights but only one real mission.

The set-up for fast-roping was very different from our normal set-up. The co-pilot seat which I normally sat in was turned around, facing into the back of the helicopter, and the co-pilot door itself was removed, so that I could just lean out of the helicopter and look down to the ground. The rear cabin door was locked open, the same way it is for a winch mission.

One end of a thick rope, anywhere from 15 to 30 metres long, was attached to the helicopter, while the rest of the rope was coiled inside the cabin, by the open door. We carried up to four STG squad members — a group of highly armed and specially trained police officers who are ready to deploy out of the helicopter.

When we reached the area where the squad was going to deploy the pilot brought the helicopter into a hover, the rope was pushed out of the helicopter, and once I confirmed that the rope was on the ground, the police started deploying. My job during the operation was to communicate with the pilot to make sure that the rope remained on the ground. If the helicopter rose and the rope was left dangling in the air then the squad members could be seriously injured when they ran out of rope before they made it to the ground.

When fast-roping there is little difference between training and real missions. The person going down the rope is not attached to the helicopter or the rope. A person falling off the rope during a training mission will be just as injured or dead as a person falling off the rope during a real mission. Every fast-roping mission is treated the same as a real one.

Once the squad was safely on the ground I had to either release the rope from the helicopter, or retrieve it back inside. In a real deployment, we would expect to release the rope, but during training operations we would usually retrieve it — a rather physical and difficult task!

MY MOST MEMORABLE police mission involved the STG squad and chasing down a murder suspect who had avoided capture for several days. Harry was the pilot and I was the crewman.

Their suspect was hiding within a large forestry block, north of Wellington. He was driving a 4WD vehicle and was assumed to be armed. Our mission was to fly the STG over the forest trying to locate the man from the air. If we found him then the STG squad was prepared to fast-rope into the area to apprehend him.

As I sat in my rear-facing seat, looking at the STG boys in the back, I could see something different in their eyes. We weren't on a training mission. We were on a search for a murder suspect; someone who was probably armed, and already proven to be very dangerous. I realised that fast-roping wasn't the only potential high-risk activity that day.

Even with the doors opened, and the wind sweeping through the helicopter, the tension was palpable. I could practically smell the testosterone in the air. With the squad all having helmets on their heads, goggles on their eyes and balaclavas covering the rest of their faces, the only thing I could see was their eyes. Their eyes were focused. They were primed and ready.

The police STG squad fast-roping onto a ferry during a training mission.

Harry headed to the area we were asked to search first. As we approached, all eyes were looking outside the helicopter, trying to spot the suspect or his vehicle through the thick bush. There were also a lot of police on the ground, searching below the tree line. Sometimes a helicopter can be effective in scaring a suspect out of their hiding position and, with any luck, into the arms of the police below.

A few minutes into our search we heard a radio call that the guy had been flushed out and had driven off. Harry headed over and less than a minute later we were above the police pursuit. The suspect was in his 4WD and was being pursued by several police cars. We joined in overhead.

The adrenaline in the back of the helicopter was now through the roof. Harry positioned the helicopter so the vehicles were on my side — this gave the police in the back the ability to watch what was going on. When I looked to my right, I could easily see the suspect who was just a couple of hundred feet below us. When he looked up at us I could look straight into his eyes.

The STG team leader asked for permission to bring their sniper to the door. After I checked his safety gear, Harry gave the permission, and the sniper moved into a sitting position, his butt in the doorway, and his feet sitting on the skids. He was sitting directly in front of me, with the toes of my boots touching his leg. His sniper rifle was close enough to me that I could have reached out and touch it.

All of this was happening while we were flying along at well over 100 kilometres an hour, doors open, wind sweeping through the cabin, police radios blaring in my ears and the STG guys wanting to be involved. It was a real chase, and he was a real murder suspect, but I never could escape the feeling that I was taking part in an action movie.

Safety is always our top priority and Harry did a great job of not getting caught up in all the drama. He managed to stay alongside the car, around the curves on the road, keeping us above the tree and wire line, and calling out what was ahead.

Normally I would be facing forward, so I could assist with some of the safety lookout stuff, but while I was facing backwards the best

I could do was swing my body around and look over my shoulder from time to time.

A few minutes later the suspect lost control of his vehicle and spun out on a corner. He jumped out of his car and started running, with police officers and a dog in hot pursuit. Seconds later they had their man in custody and it was all over. The police got their man!

We flew the STG squad back to their vehicles in the forest, and then headed to the airport to get ready for our next mission. Another job well done.

OVER THE YEARS we would debate whether or not a rescue helicopter should get involved in police missions. In the end we decided that protecting public safety was an important part of what we did, and if we could assist the police or bomb disposal unit, we would be taking a proactive stance in saving lives.

The police missions were few and far between, but they certainly make up some of my more memorable missions!

Unfortunately, much of what I saw and heard cannot be written about. One of the cops told me that if I revealed their secrets they'd have to shoot me. I'm pretty sure he was kidding, but why take the chance!

THIRTY-NINE
MY LONGEST AND MOST DIFFICULT FLIGHT

LIVING HALF A world away from my family was always going to have its challenges, and as my parents aged I knew that one day I would get the call we all dread. It took 22 years, but in February 2012 it finally came.

It was the morning of my first day back at work after a week away and we were getting ready for an inter-hospital transfer of two pregnant women from Masterton to Wellington. A few minutes before we took off, my older sister Linda called to tell me Mum, who now lives in Florida, was unwell.

After running tests, doctors had found that Mum's heart was beating too slowly and she needed a pacemaker to regulate her heart rate. Normally, this was fairly straightforward, but further tests revealed that she also had a leaky heart valve and an aortic aneurysm. Straightforward had become life-threatening. Unlike Doug's dad back in 2001, my mum's aneurysm was not leaking. Even so, it can be a risky operation.

The problem with having medical knowledge is that I knew the worst-case scenario, but I took it as a good sign that they had not rushed her straight to theatre for urgent surgery; it was scheduled for two days' time. It was still extremely serious, however, and I wanted to get there as soon as I could to support her and Dad.

Both of my sisters lived in New York and they had booked flights for the next morning. I told Linda I was about to go flying but I would book the first available flight to the US, which I knew wouldn't be until that evening, and arrive in Florida before Mum had surgery.

I needed to arrange coverage for my upcoming shifts, a big ask seeing that I had just returned from leave, but I knew that my team would do whatever was needed to keep the helicopter flying and to help any of us out.

There was nothing I could do then to change things for Mum and no way for me to get to her any sooner. The best thing I could do right now was go on this helicopter flight.

On my way to Masterton I managed to book flights to Miami, about an hour's drive from my parents' house. My flight from Wellington would leave around 5pm so I had plenty of time to complete this mission, throw a few things in a bag and get to the airport.

Once that was sorted, I called my boss to tell him I had to head to the USA that afternoon, and I wasn't sure how long I'd be gone. Then I rang Colin, the assistant operations manager, who said he'd take over my shift when we got back to Wellington and that he would arrange cover while I was away. Everything came together quickly and I was grateful to work for a caring organisation which understood that emergencies happen and came together to look after each other so well.

Once we safely delivered the two patients to Wellington Hospital, I finished off my paperwork and headed home. I gave my dad a call and he was, as expected, upset and worried about Mum. He was happy to hear I was coming.

My good friend Bettina and her daughters took me to the airport and kept me entertained while I waited for my flight to Auckland. It was nearly 12 hours since my sister's call by the time I got on the flight to Los Angeles, and I still had another 22 hours before I would reach my parents' house. I was already physically and mentally exhausted.

Just before we took off, I checked my phone and email one last time. It was now the middle of the night in Florida so I was relieved that there were no updates. At this point no news was good news!

Up to that point, I had managed to keep myself distracted, but now I had nothing to do except take stock of what was happening. Mum was in her seventies and in reasonable health, but was about to undergo major surgery, where the odds were not necessarily on her side. However, she is a fighter and I was sure that she was mentally strong enough to cope with it and the long recovery. I was more worried about Dad; my parents had been together for over 50 years and I knew how much they loved each other. I was unsure how Dad would handle it, especially if things didn't turn out well.

The flight to Los Angeles was, without doubt, the longest 12 hours of my life. I managed to get a few hours' sleep but it was a horrible flight filled with a sense of helplessness because there was nothing I could do. It was frustrating being out of touch and not knowing what was going on. As soon as the plane touched down, I texted and got an immediate reply — Mum was still okay. I was pleased, but it was still another 10 hours before I reached my parents' house in Florida.

I finally pulled into their driveway at around 1am, 35 hours since I first got the news. Dad was asleep but my sisters were awake and waiting up for me. After some tears and hugs all round, I collapsed into bed for a few hours' sleep. We were all up around 6am and off to the hospital a little while later and I was relieved that Dad seemed to be coping well, in fact better than I expected.

Normally, the distinctive sights, smells and sounds of a hospital are comforting to me, but today, as a distraught family member, it made me uncomfortable. We got to Mum's room as she was about to head to theatre. She was delighted to see me and I managed to hold it together as I gave her a kiss and told her to stay strong. As the doors closed behind her, I stood there with tears in my eyes, wondering if I would ever see her again.

The four of us had a restless day, telling stories and sharing memories, waiting for word on Mum. It was early afternoon before her surgeon came out and told us that the surgery went well and we could visit her in the ICU soon. I made a light-hearted joke about not needing my suit for the funeral and discovered my family didn't share my dark sense of humour. They could not believe I had packed a suit. It seemed I was the only practical one in the group!

Mum's breathing tube was removed the next morning and she started the long, slow recovery process. I am sure I get my sense of humour and practicality from her, because when I told her I had brought my suit, she told me not to waste money on getting it dry cleaned!

A couple of days later my sisters headed back to New York. I extended my trip so that I could stick around for an extra week to give Dad a hand and spend time with him and Mum. This was the first time that he and I had spent so much time together in over 30 years, and although it was tense at times, it was a special time for us. We had a chance to talk about the good and bad of my teenage years and how much our views had changed over time; he was proud of what I had done with my life.

We visited several different heart rehabilitation units together, trying to find the one that would be best for Mum. My criteria were weighted towards the medical side, and his to the little things they could do to make her more comfortable; in the end we found a place we both thought offered just the right mix.

A few days later, Mum was discharged to the rehabilitation unit and it was time for me to head home. Fortunately, Arlene and her family had already planned to visit for a holiday in Florida and they arrived the morning I left, which meant my parents had ongoing support.

For the first time since I had moved to New Zealand I found it hard to live so far apart from my family. My friends, the job I loved and my life were all in Wellington but I resolved to find a way to spend more time with my parents.

I promised myself that from now on, I would visit my parents at least twice a year, a promise which I have kept. After spending so much time in ICU as the relative of a patient, I also headed home with more insight than before. As much as the ICU nurses looked after Mum, they also looked after us distraught family members.

We often don't understand something until we experience it

ourselves, and it definitely made me more empathetic and understanding towards the people I saw daily on my job.

THE FOLLOWING YEAR, in July 2013, my parents celebrated their fifty-fifth wedding anniversary on the same day that their grandson Jason, my nephew, married Jessica, the love of his life. It was very special that the two couples shared a wedding anniversary. A couple of days after the wedding, I drove with my parents from New York to Florida. I think we were all nervous about spending that much time locked in a car together but it ended up being a lovely 20-hour drive, spread over a few days.

Being trapped in a car with me, they had nowhere to run when I asked about their health and the medications they both took. Conversely, I had nowhere to run when they started asking me about my love life! I was keeping myself occupied, and my sisters amused, with a constant stream of texts that I sent when I was not driving. From my point of view, the funniest one was: 'According to Dad everyone in North Carolina are assholes. The people we pass are slow assholes, the ones that pass us are speeding assholes.'

During that drive, I also posted a Facebook photo of them, holding hands as they walked towards a restaurant. The caption read 'A surprisingly pleasant 20 hours on the road with my parents. Safely at their home in Florida, sanity intact. And here is a cute picture showing how much they love each other days after their fifty-fifth anniversary.'

If there was one thing that was always on show, it was my parents' love and affection for each other.

During one of my regular visits in April 2015, it was apparent that dad was unwell so I made him move a scheduled visit with his cardiologist forward, to the next day. In scarily similar circumstances to my mum, the doctors discovered that he had heart problems and needed similar surgery to her. At least this time I didn't have to make a mad dash across the world to be at his bedside. I picked my sisters up from the airport late that night, and the next morning Mum, my sisters and I were there as Dad was rolled towards theatre.

Dad survived his surgery, and things were looking as positive for his recovery as they had for Mum. By the time I headed back to New Zealand about ten days later, Dad was making a good recovery and was looking as fit and healthy as could be expected. After I left, both my sisters made quick trips to Florida to keep an eye on the folks and things were looking good.

Unfortunately, Dad took a turn for the worse a couple of weeks later and ended up back in ICU. Mum seemed to be coping well with it, but I was relieved to hear that Arlene was heading down for a visit. She visited with Dad on the afternoon she arrived and he was pleased to see her. However, when she rang me, she was concerned about how much he had aged in the few weeks since she last saw him.

That night I was babysitting Zara and Georgia, and around 6.30pm while I was making their dinner, I got a call from Arlene to say that Dad had passed away in his sleep. It was upsetting to hear, but not totally unexpected after her call earlier in the day.

Dave and his sisters, Linda and Arlene.

While I was talking with Arlene, Georgia who was eight at the time, hugged and cried along with me while she listened in on the call. I let the girls eat dinner in front of the TV, which was never allowed, so Zara asked me why. I said she could thank my dad, and with all the innocence of a five-year-old she responded: 'No we can't, he's dead.'

I arranged to head to New York for the funeral, but I didn't have time to reach Auckland that night to meet a US flight so I planned to leave the next day. Again, my wonderful friends were by my side to support me that night. Bettina and Grant came back home to comfort me, and offered to cancel their night out, which was kind but unnecessary, and then Doug came around and we drank a nice bottle of red in memory of Dad.

I had tucked the girls into bed a bit earlier and when I went up to check on them Georgia had left me a note, which reminded me how much love and support I had right here in New Zealand.

> I love you dave and I am sorry about your dad. I love you so much. Love g you

THE TRIP BACK to New York was less stressful but sadder than when I rushed home for Mum's surgery. I arrived in New York just after midnight Friday morning, we buried Dad that day and I was back on a plane, heading home on Monday afternoon.

Apart from my grief for Dad, I was also worried about Mum being on her own, but I knew that she was strong enough to cope. I was so happy with my decision to visit my parents twice a year because it meant that I got to spend time with Dad while he was still healthy. I didn't have regrets about things I should have said or done because I was spending so much quality time with them.

I was also reminded about the love and support I have in New Zealand. Family are not just the people you are related to.

FORTY
WHEN THINGS GO WRONG — A FEW CLOSE CALLS

THERE WAS NO doubt we weren't making it to land — we were going down into the cold water below. As my mind ran through my emergency procedures one last time I pulled my seatbelt as tight as it would go, planted my feet firmly on the floor, tightly gripped the bottom of my seat with one hand. I placed my other arm across my body and grabbed the opposite shoulder. The last words I heard were 'BRACE! BRACE! BRACE!' and then we dropped into the water.

Immediately after we hit, the helicopter began turning upside down and the cabin started filling with icy cold water. As I felt the water reach my face I took my last, large breath of air and waited for the wreckage to settle. Suddenly, everything became strangely quiet.

I could not get the door next to me open so I knew I would have to make my way across the cabin to escape through the other door. I was disorientated by the total blackness and the fact we were upside down. When I had my bearings, I removed my seatbelt and immediately felt myself being pushed upwards, towards the cabin floor, which was now above me.

I pulled myself across the cabin, getting caught up by the myriad cables floating around inside it. Suddenly I was kicked in the face and chest by flailing legs in front of me. I couldn't see her, but I knew this had to be Tania, the flight nurse who was in the rear cabin with me. I was glad she was working so hard to survive. I waited for her to swim out the window and then followed her through.

I reached the surface and took a deep breath. I removed my darkened goggles and I could see Tania a few feet away, grinning like the cat that got the cream.

We had both survived our latest round of HUET, Helicopter Underwater Escape Training.

JUST ABOUT EVERY helicopter flight from Wellington Airport involves flying over a body of water, so being prepared if things go wrong is essential. The New Zealand Air Ambulance standard requires anyone who flies on a rescue helicopter regularly to do a HUET course every two years. Not everyone is a fan of the course but personally I love it. In the highly unlikely event that we ever crashed into the sea, I knew my chances of survival were greatly increased, thanks to the training.

Flight helmets, flame-retardant flight suits, a survival vest, boots and cotton under-garments were all part of my standard uniform; any of these bits could help me survive a crash.

A major part of my emergency service life has been being prepared for the 'what if' scenario. For a group of people who didn't expect to crash, we spent a lot of time and money to ensure our teams were trained and prepared for when things go wrong.

Some of the safety gear we wore or carried in the helicopter.

Colin and Julian preparing to 'crash' in a HUET simulator. *Hannah Latta*

A safety diver keeps an eye on Colin and Julian as the 'helicopter' hits the water. *Hannah Latta*

MOST DAYS, I FELT safer in a helicopter than I ever did walking through Wellington and crossing Lambton Quay. Throughout my career, I've been lucky to fly with the best pilots in the world in well-maintained helicopters alongside crew and medical staff who always had each other's back. Vitally, we had the training and equipment needed to ensure we stayed safe. So, what could go wrong? Plenty!

Everything in life, including crossing the street, can be risky. Hanging off a moving helicopter is clearly more dangerous than working in an office. Even staying in bed carries some risk with it. The trick is to manage risk, not to avoid it.

I believe that we learn more from things that go wrong than we do from those that go right. I was taught early on that these are great 'learning opportunities' and only failures if they kill you or you keep repeating the mistake.

I have had many learning opportunities over the years — those I've created and those I've watched others create — and I've tried to learn from each of them. Although learning opportunities happen every day, I strongly believe it's not a lesson learned until behaviours change.

I've been incredibly lucky that my worst on-the-job injuries have been cuts, bruises and a dog bite. My longest-lasting injuries are a continually bruised ego from all the learning opportunities I created for myself.

SANTA CLAUS

For many years, Life Flight would take part in Wellington's biggest Christmas celebration, 'Carols by Candlelight' on the city's waterfront. Each year we would hover over the Harbour near the event and winch Santa, and sometimes an elf, out of the helicopter to wave to the 10,000 people assembled in Frank Kitts Park.

Some years I was Santa and some years I was the winch operator, but I always managed to take part because this was my favourite PR event of the year. For the 2001 event, as I was working a nightshift, I was the winch operator. The two people wearing the Santa and elf suits came in early for a briefing.

We had completed our winch check and were sitting on the tarmac, ready to start up, when my phone rang. The ICU team had a patient they needed to retrieve from Masterton, but they needed 30 to 40 minutes before they'd be ready to leave. I had a quick chat with the pilot Grant and we agreed that there was enough time to do our Santa appearance and then set up for the emergency flight.

To save time when we got back, I disconnected my safety strop, jumped out of the helicopter and went into the hangar to get some gear prepared, while Grant started up. When we were ready to take off I settled myself back into my seat in the doorway.

During the short hop to the city, my mind was split between the upcoming winch operation and the upcoming flight to Masterton. As the lights of Frank Kitts Park came into view, I focused on the winch; it didn't matter if it was a PR event or a real rescue, suspending people off a hovering helicopter needed my full attention.

I moved Santa and his elf into the doorway and then stepped out on the skid as we winched them about 10 metres below the skid. A large spotlight from the stage lit up Santa, and we swept our 30-million candle-power Nitesun across the waving crowd. It was as

Dave warming up as Santa.

252 Emergency Response

much fun as ever but I was anxious to finish up and get back so that we were ready on time for the ICU transfer.

Once Santa and his elf were safely on board the helicopter, I waved to the crowd as we turned and headed for base. It was a beautiful night so I stayed in the doorway while we flew back to the airport.

After we landed, I reached behind me to disconnect my safety harness from the strop connected to the helicopter and, to my absolute horror, discovered that I was not connected!

I had just sat in the doorway as we flew 300 metres (1000 feet) above the ground *and* I had stood out on the skid 30 metres (100 feet) above the water, and at no point was I actually connected to the helicopter! One slip and I would have fallen to a certain death.

The phone call from ICU had broken my focus for the job at hand. When the phone call came in, I had already done my mental and physical safety checks. Disconnecting myself from the helicopter was not part of the routine, and when I jumped back into the helicopter, my mind was focused on the 'real mission' and I simply forgot to reconnect myself.

Looking back, I would rate this as my biggest screw-up. One moment of losing focus could have killed me. Thankfully, it didn't, but I did learn from it. From that day on, I gave my safety strop a good tug at both ends: where it connected to the helicopter and where it connected to my body.

OPEN COWL

My first in-flight scare came on a dark winter's night in 1993. We were overhead the scene of a fishing boat which had sunk in the middle of Cook Strait. We were 10 kilometres from land when a vibration was felt in the aircraft accompanied by a horrible banging noise. I had to admit, it was terrifying but my nerves were tempered by the calmness of Toby, who simply headed back to shore and reassured me that everything felt okay in his controls.

Once back at base, we discovered that a cowl, the compartment door over the left-hand engine, had not been properly latched and had opened during flight. The rotor blade continuously striking the cowl caused the noise and vibration.

We got lucky; the unlatched cowl was on the same side as our tail rotor and if it had ripped off and struck the tail rotor, it would have likely sent us into a deadly spin over the cold and deep waters of Cook Strait.

How and why it happened were less important than the safety improvements we made. The helicopter was fitted with a warning light which alerted those inside if a cowl was not secure. More importantly, for me, I never again let the urgency of a mission override the need for a thorough walk-around safety check before every flight!

LOST REFERENCE

In 2002, I had another big fright on a cold and dark winter's night. This time we were responding to an emergency beacon which had been activated for a man suffering from heart problems in a hut in the Richmond Ranges, near Nelson.

Brian Taylor was the pilot, Dean the paramedic and I was crew. We arrived in the area and Brian brought us through a large hole in the clouds, tracking the signal down towards the valley where the hut was located. The cloud was broken, giving us enough moon and starlight to use night vision goggles (NVG). However, this was still in the bad old days when our pilots couldn't use them, so Brian was reliant on external references.

We had descended to tree-top level and were slowly making our way towards the hut when Brian lost his visual reference of the trees on his side of the helicopter. For him it was like staring into a big black hole, with no idea where the helicopter lay in relation to the trees or the ground, or the angle of the helicopter. It's a bit like waking up in an unknown room, in complete darkness, and having to walk a tightrope to find the door.

Dean and I, who were on NVG, suddenly saw that we were flying backwards at a nose-up altitude, instead of forward towards the hut.

Fortunately, Brian was such an experienced pilot, with thousands of hours of instrument flying, that he instantaneously flicked into helicopter instrument flying mode and worked quickly to regain straight and steady flight. Meanwhile Dean and I talked Brian through the scene, creating a clear visual picture for him of our spatial location. This is what all those hours sitting in a car talking the blindfolded driver around the airport tarmac were for.

It only lasted five or ten seconds, but they were amongst the most terrifying of my life. I may have been outwardly calm but I was sweating profusely and my heart was pounding by the time we were back in a steady hover. A quick check and we all agreed we were okay and since we were now practically overhead the hut, continued with the rescue.

This rescue highlighted how important excellent CRM was, and it was also another catalyst to push the Civil Aviation Authority into allowing pilots to fly using night vision goggles. That took a few more years, but it did happen.

WHERE THERE IS SMOKE . . .

The last in-flight emergency I was involved in was around 4.30am on yet another winter's morning in 2009. Mike was flying and we had a neonatal team on board heading to pick up a sick baby from Palmerston North. The helicopter heater was working overtime, trying to keep us warm.

We were about 30 minutes into the trip, flying at around 1500 metres (5000 feet), when Mike and I both thought we could smell smoke. The smell quickly went away so we assumed it was the heating system, which often generated a bit of a burning smell. A few minutes later the smell was back and this time acidic smoke started to build in the cockpit. Mike immediately descended towards the farm land below and called a 'Pan Pan', one step below a 'Mayday' to let air traffic control know we had a problem. Meantime, I briefed the neonatal team about the emergency procedures and reminded them that as soon as we landed they had to get out of the helicopter and move as far away as possible.

The smoke got thicker in the cockpit and at one point I told Mike that I could feel heat under my seat. We discussed popping our doors off to get fresh air into the cabin. As it was still dark, we had to quickly identify a landing site, using the moving map, our NVG and the Nitesun to check for wires or other obstructions between us and the paddock we were heading for. Why did these things always happen at night?!

As soon as we touched down, the doctor and nurse didn't need any encouragement to jump out and head away from the helicopter. While Mike shut down, I tried to locate the source of the fire. It turned out that a computer circuit board had shorted in the nose of the helicopter, creating the smoke. The heat under my seat must have just been my butt cheeks clenching!

A road ambulance picked up the neonatal team and their incubator, and they continued to Palmerston North. After removing the offending circuit board and getting an okay from engineers, we returned to Wellington.

Although in this instance it wasn't serious, fire aboard aircraft can be deadly, so it was reassuring to see how well we performed — our calm approach to the emergency, great flying skills from Mike and solid teamwork kept everyone safe.

FALLING PARAMEDIC

In the early 1990s, after a report about a fatal winch accident, we implemented new procedures for boat winching so that when we winched from a boat, we would lift the people off the deck and then move over the water and descend, just in case anything went wrong. This change in procedure proved to be a life saver during a mission in November 2000.

We departed Wellington at 6am to retrieve a sick worker off a commercial fishing boat 260 kilometres southeast of Wellington. Hugo was flying and we had a WFA winch-trained paramedic with us, who was scheduled to finish his 14-hour shift at 7am.

We winched the paramedic onto the boat and he disappeared below deck to check on the patient. A few minutes later the paramedic and patient were on deck and ready to be winched to the helicopter. As we lifted them off the deck, we moved over the sea and Hugo descended to keep them about 10 metres above the surface.

The paramedic appeared to be struggling and then, to my utter dismay, he fell off the winch hook and into the sea below. I continued to winch in the patient while I kept an eye on the paramedic in the water. I told Hugo what had happened and he moved back towards the paramedic while I brought the patient on board. The fishing boat's captain had also seen what happened and I watched his 80-metre-long ship executing an emergency turn while people on deck scrambled to keep the paramedic in sight. Seconds later we were back overhead the paramedic, who was floating on the surface and gave a wave to let me know he was okay, to my huge relief.

We winched him up and he was okay. He told us that he was tired and began to feel seasick when he got on the ship. Below deck, he was hit by smells of diesel and fish guts, which made him worse. Tired and sick people lose focus so when he was about to be winched off the deck, he missed getting his carabiner on the hook. As soon as he lifted off, he realised his mistake but instead of letting go and sending the patient up on his own, he held onto the hook, thinking he could make it to the helicopter.

We were relieved he was okay, thankful that there had not been a gear failure, and extremely happy that we had followed procedures. Had he fallen onto the steel deck of the ship instead of into the sea it could have killed him. Instead his worst injury was a bruised ego and a dunk in the ocean.

It also highlighted the risk of using any team member when they are fatigued. We changed our procedures to avoid using paramedics at the end of their shift and added some new training and additional safety gear to make it easier to secure a winch hook to a paramedic harness.

THESE ARE ONLY a few of the many 'lessons learned' over my career. The fact that we updated procedures, enhanced training and brought in different equipment are all signs of organisations working to minimise risk.

I was always grateful that the safety culture at Life Flight was strong, and that people felt they could report accidents and mistakes so that others could learn from them. We worked in an environment full of risks and one of the best ways to minimise and eliminate them was to talk openly and share the lessons learned.

FORTY-ONE
MANAGEMENT TEAM

LIFE FLIGHT HAS grown thanks to the hard work of many people, on the aircraft and behind the scenes, over the years.

All the helicopter crew, including John and me, were unpaid volunteers until 1995 when the Accident Compensation Corporation allocated additional funding on a trial basis for the service. ACC's special funding ended a year later, but the Trust saw the value of having full-time crew, so the paid positions remained. Around the same time, the fixed-wing service began paying its crew on a casual basis, which continued until Phil was brought on as the first full-time fixed-wing crewperson in 2000.

While the flying has always been the best and most important part of my job at Life Flight, there has been a lot of the behind-the-scenes development work and the building of strong relationships with key stakeholders that has made an enormous difference to the organisation.

As a small charitable trust, everyone who joined the Life Flight team was expected to add value in as many ways as they could. We all brought different skills into the mix and did what we could to help in other areas.

One of the agreements that John and I made in the early days was he would do screwdriver stuff and I would do computer stuff. While he designed and built the equipment we used in the aircraft, I wrote a program to collect mission data and collect statistics which could be used by the fundraising team. The agreement worked well and continued for all the years we worked together!

I am very proud of the things that I helped create to make Life Flight the great organisation it is today.

I became part of the management team when I was appointed as Crew Chief in 1995. I was John's assistant, looking after some of the administration of operations. My IT skills came in handy as I continually updated our mission database and set-up and maintained our growing network of computers. In 1999, I wrote the ISO manual that still forms a major part of Life Flight's policies and procedures today.

Some of our toughest days as a management team were in January 2003, when the helicopter hit trees while heading towards Masterton Hospital. Thankfully, everyone on board walked away from the crash, but it certainly tested our resilience that night, and in the days and weeks after. I learned a lot about being a good leader by watching Kevin, our General Manager, at the time.

When John stepped down as Operations Manager in 2005, I took up the role until it was disestablished in 2014. As Crew Chief and Operations Manager, I was part of the management team which saw the Trust through massive growth in the number of patients transported.

Managing teams that work 24/7 and are constantly dealing with death, trauma and other tragedies, brought unique management challenges for John, Ian Lauder (our Auckland Operations Manager) and me. Anytime a team member called in sick, or was otherwise unavailable for a shift, we had to arrange cover on short notice, otherwise our aircraft would be grounded. We also had to be available 24/7 to help support our teams who were involved in stressful or emotionally draining missions.

Working with my colleagues on the senior management team was amongst some of the highlights of my career. As a team, we worked well together and helped move Life Flight from strength to strength and kept the organisation running smoothly.

I believe one of our best accomplishments, was reaching an agreement with Wellington Free Ambulance in 2010 to put full-time winch-trained paramedics on the helicopter.

OVER THE YEARS, I developed five golden rules of management that I held myself to, both as a manager and when I was on the helicopter. Whether we were looking at spending money, introducing a new technique or deciding whether to undertake a flight, I would assess everything against these rules:

1. DOES IT IMPROVE OUR TEAM'S SAFETY OR ABILITY TO DO OUR JOB? The health and safety of our team was always our number one priority.

2. IS IT GOOD FOR THE PATIENT/PERSON WE ARE TRYING TO RESCUE? At the end of the day, if we were going to spend a lot of money on a new widget, then the widget should improve our abilities to help save a life.

3. HOW WILL WHATEVER WE ARE ABOUT TO DO LOOK ON THE FRONT PAGE OF THE *DOMINION POST*? The 'front page' test is a well-known and practised test. We were always lucky that our public perception was high, and good, but one scandal, or terrible decision, always had the ability to change that perception.

4. WILL THIS END UP WITH THE BOSS SAYING, 'YOU DID WHAT?!' Of all the rules, this was probably my go-to. Trying to explain a decision to the CEO, particularly one that does not work out the way you hoped or intended, is a great moderator. I would imagine myself standing in front of him or her, explaining why I did or didn't do something, watching their expression change, their blood pressure rise and then, finally, them standing up and shouting: 'YOU DID WHAT?!'

5. WE ALL SCREW UP, JUST TRY NOT TO REPEAT THE SAME MISTAKE TWICE. Even better, try to learn from someone else's mistake, and don't repeat that.

These rules would hold up well in most organisations. When I consult with senior management teams today, I am able to give solid examples of how to apply them within their own organisation.

SOME OF MY best management training happened working with rugby teams. I spent ten years working as a volunteer team liaison for teams visiting for the Wellington leg of the World Rugby Sevens Series. As liaison, I would work with the team manager to ensure the team always had what they needed, and got where they needed to be on time. Each team was assigned two vans which we would use to drive them around. Over the years, I worked with the teams from USA, Fiji and Argentina.

The Fijians showed me what being on 'Island time' meant, as they were always a few minutes late when we were departing for training or heading to a PR event. I fell into the trap of believing they would always operate like this, until the evening we were scheduled to visit the Fijian ambassador's residence for dinner. We were supposed to depart at 6.45pm, which I assumed meant closer to 7pm. As we

Dave with the 2011 Argentinian Sevens team.

262 Emergency Response

pulled our two vans up to the front of the hotel at around 6.40pm, I was shocked to see the entire team waiting. The manager explained that 'Island time' was a great way of life, but when they had important appointments, like ambassador visits or getting to the stadium for a match, they'd always be early.

Thanks in large part to my Sevens experience, I was selected as one of the team liaisons for the Rugby World Cup, looking after the Namibian team. Playing an important role in one of the biggest events to ever hit New Zealand was fantastic, and it tested my management and logistic skills to the max. Working with their management team to keep 45 people happy as we travelled around New Zealand was an exceptional experience. The night I took six of their players to a strip club, to keep them safe, rates highly in my book of good management decisions, but that is a story best left untold here!

I've also had the opportunity to be the team liaison for the Argentinian Pumas and the South African Springboks when they played a test against the All Blacks in Wellington. Spending time with elite athletes, watching the way they trained, ate and acted with each other, was a pure demonstration of how to be at the top of your game, no matter what the game is.

Ruth Zeinert sitting in the doorway of the helicopter as it lands in Westpac Stadium prior to a rugby match.

I WAS ALSO fortunate to work on many external committees, including representing the Air Ambulance sector on the committees that wrote the first New Zealand Air Ambulance and Air Rescue Standard and the first version of AMPLANZ, the Ambulance sector's response to major incidents and crises.

While the operation teams in Auckland and Wellington were helping a record number of people on the aircraft, in the background the administration and marketing teams kept the money coming in and the bills paid. Running aircraft is an expensive business, and these teams did an incredible job of generating community support and sponsorship.

I AM, AND always will be, proud of my work on the management team, but if I had it all to do over again, there are three reasons why I would choose not to join the management team.

The first, if I was not a manager, then I wouldn't have been involved in the review that led to the disestablishment of my position. I'd probably still be a full-time crewman today.

The second, and more important reason, was that it drove a wedge between me and the people I worked with on the aircraft. One of the things I was warned about in my first management course was that it is impossible to be someone's friend and their manager.

It also meant I was at odds with some of the pilots from time to time because we worked for different organisations with different wants and needs. The thing that I valued most is that we always set our differences apart and formed a cohesive team on the helicopter, where it mattered most.

FORTY-TWO
MANAGEMENT REVIEW — THE BEGINNING OF THE END

LIFE FLIGHT WAS more than a job to me, it was my passion. I often joked that we would eventually need a bigger helicopter so my Zimmer frame could fit in it when I got older. I really believed that I would be part of Life Flight, in some capacity, forever.

My intention had always been to fly for as long as I was physically fit, or until I couldn't handle the shift work any more, and then continue in some management position, or as a trustee, for many years to come.

The members of our management team were surprised when, in late 2013, the Board of Trustees announced that they had contracted a consultant to undertake a management team review.

Over a few weeks, each of the management team met one-on-one with the consultant. After we met, none of us was particularly happy because we didn't feel that our individual meetings were comfortable or productive. During my meeting, I felt like there was an agenda that I knew nothing about and I walked out dejected.

The week before Christmas, the trustees called us together to discuss the results of the review. When we walked into the room, an ominous stack of envelopes and papers sat on a table. A feeling of dread hit me and when I glanced at one of the other managers, they looked like I felt.

The Chairman thanked us for coming and started to read through the report. It quickly became clear that changes were coming. Then I heard the words which changed my life forever: 'It is proposed that the Wellington Operations Manager position be disestablished, to be replaced by a new position.'

I suddenly found myself numb, in a state of shock and trying to process what I had just heard.

I tried to keep it together while the meeting dragged on, but I started to feel physically ill and needed to get out of there. By the end of the meeting, we all seemed shell-shocked by some of the changes proposed. My colleagues were also my friends, and they all tried to find something comforting to say to me as we left, but no such words existed.

I excused myself from the group, jumped in a lift and made my way to the ground floor of the office building. I was in a daze as I walked out onto the street, unsure of where to go, who to call or what to do. The office building was on Lambton Quay, one of Wellington's busiest streets, and I was startled into reality when I heard the blaring horn of a slow-moving bus as I absent-mindedly walked in front of it.

In the space of an hour, I had metaphorically been thrown under a bus and was almost hit by a real one. I had to get off the street and find a friend — quickly. I called Doug, who worked only a couple of blocks away and told him enough so that he understood what was going on. We agreed to meet at a nearby café, one where I didn't have to cross any more streets! As I walked there, I called Bettina, and told her the shocking news. She was as stunned as I had been and we agreed to meet up a bit later.

Doug must've run the whole way to the café because he arrived within minutes. He let me say whatever it was that came out of my mouth, mostly babble. The two of us often share our problems with each other and, being boys, offer each other solutions. Fortunately, he realised that today he just had to let me talk, or cry or say whatever came into my mind, and he did this well.

While we were talking I suddenly switched from being hurt and upset to being extremely angry. When this happened, I started to feel the blood flow through my body again. Doug said he watched the colour return to my face and some spirit return to my heart. I was hurt and devastated, but this was just a proposal and I was going to fight to stay in the job I loved.

I spent the rest of the day talking with my family in the USA and surrounded by my friends in Wellington. Word got out quickly amongst my closest friends and they were all there to support me in any way they could.

Originally, the proposal document was to be released to all staff the following day, with feedback due by Christmas Eve, which all the managers strongly objected to. Finally, it was agreed that staff wouldn't be given the proposal until mid-January, after the Christmas break.

My CEO knew that I was upset and offered to stand me down from my flying duties, if I needed a break. However, the helicopter missions were more important to me than ever and I knew I was focused and could put everything else out of my mind when I was on a mission. Even so, I sat down with my pilots and let them know what was going on because safety is paramount and they deserved to know that the possibility existed that I could be distracted on a job.

I had a beer with Kevin, a good friend who had been my General Manager for about seven years, and knew me well as a person and as an employee. He reminded me of many of my accomplishments and the good things I had done for the Trust, over my career. He summed it up by saying that I had worked hard to build up a legacy over the past 23 years, and I was the only one who could destroy that legacy. He advised me to keep my head high, accept whatever decisions were made, and if I did lose my job, not try to hurt the organisation I loved or go out kicking and screaming.

This was the single best piece of advice I got from anyone and it formed the basis of my thinking and actions from that day on.

Even though it wouldn't be officially announced for weeks, I began confiding in my wider network, including some of my co-workers. Without exception, anyone I told about the proposed restructure was shocked and didn't believe that the Trust would go through with it.

I had been in the room when the plan was announced, and didn't share their level of optimism.

If there is one thing I know about myself it is that I am resilient. No matter what the outcome of the employment process, I had my health, I had my brains and I was surrounded by people who loved me and cared about me. I knew I would be okay. I decided I would fight like hell for my job, but if I didn't manage to keep it I would find a way to get through this and come out the other side a better person.

The thing I struggled with most is that I could not get a simple question answered by the Trust: if I lost my management job, would they keep me on as a crewperson? I had one employment contract which included all my roles so if I lost the manager role, my flying role went with it. Being part of the rescue crew was important to me and I wanted to keep flying. It was incredibly distressing not to know.

After the restructure proposal was announced to staff in January, news of it hit social media, and then the mainstream media. I was surprised that what was essentially an employment matter ended up in the news.

The messages I received from my friends, colleagues, people I had helped rescue and complete strangers were humbling and overwhelming. Even if the Trust didn't want me, it was clear that I had huge support in the wider community.

I put together my best arguments and alternatives to the Trust's proposal, but all to no avail. On 12 February 2014, my lawyer and I attended a meeting to hear the result of the final restructure.

As I expected, my position was disestablished.

FORTY-THREE
END OF MY FLYING

IT WAS NOW confirmed. I didn't like it, and I didn't understand, but at least now the decision was final and there was no more wondering. The Trust had plans, and they didn't include me.

My only question on the day was whether I would be able to stay on as a crewperson. They told me they couldn't give me an answer at this point, which I found frustrating, but given my recent experience, unsurprising.

While I was calm and resigned to my fate, other people seemed more surprised and distraught by the news. People at Life Flight seemed to think that some common sense would prevail, but it was not to be.

I worked with my lawyers to negotiate my exit, and it was eventually agreed that my last day as Operations Manager would be on 31 March, just six weeks away.

Once I signed that agreement, the Trust finally told me that they would like to keep me on as a casual crewperson. My lawyer explained to me that they probably had not been able to confirm my flying role until now due to the complexities of employment law.

I was thrilled. My flying was not over. We negotiated a casual contract and, for now, I would be used to fill the shifts I had been previously flying.

On my last day, I gave my final handover to Colin, my Assistant Operations Manager, who was now the Acting Manager of Operations. When we finished our meeting, I cleared out my last few odds and

Dave in happier times.

ends from the desk I had been sitting at every work day for the last 11 years. It was a sad day, the end of an era.

A couple of days later I headed off to the USA for a month-long break. I needed a bit of time to let it sink in, and plan the rest of my life.

When I returned, I began working on my business in between my flying shifts. I was picking up a lot of overtime, so, ironically, I was doing more shifts than I had been doing before. I also began covering some fixed-wing shifts, which I enjoyed immensely.

MY NEW ROUTINE was working well, and I felt safe and secure with my flying, until the end of 2014, when I heard that they had begun recruiting for a new full-time crew person. I didn't put in an application because in the past, any casual staff would be offered the chance to move to full-time, and since I was not offered the opportunity, it was clear to me I was not wanted. I was gutted, and saw this as the next step towards being pushed out the door for good.

Early in 2015, a new crewperson was hired, and after his training was completed, I was moved off the flight roster and told I would only be used to cover overtime shifts, meaning that my flying hours were greatly reduced.

I was only guaranteed one shift every 90 days, which I was told would keep me legal. My response was that one shift every 90 days might keep me legal, but it wouldn't keep me current. Regular flying, which inevitably would include winching, was essential to keep skills sharp, and one shift every 90 days was not enough to do that.

Over the next few months, I only did a few helicopter shifts each month and there could be weeks between my operating the winch, either in training or on a real mission. I was getting very concerned about my ability to maintain my winch currency and skills and shared this with my managers and co-workers.

A few weeks before Christmas in 2015, my worst fears came true. We were carrying out some winch training with the Coastguard, which involved winching the paramedic onto a small boat. It had been six months since my last boat winching and I was very rusty. The exercise didn't go well, resulting in the paramedic unexpectedly being pulled off the back of the boat and into the Harbour. She was okay, but I knew that was by pure luck; she could have been seriously injured or killed.

When I reviewed the winch-cam footage, I was horrified to hear how bad my winch patter had been. When things started to go wrong I reacted slowly, could not accurately convey what was going on to the pilot, and I had not actually said some of the things I thought I did.

The incident resulted in a big safety review and, to be fair, the CEO agreed that my lack of currency was a big contributing factor. He agreed to find a way to increase the number of shifts I was allocated.

I was very shaken up by the incident and lost a lot of self-confidence. I talked with team members and, for the most part, they told me to shake it off, that it was just one incident in a long career.

My biggest worry was that winching is inherently dangerous and

I needed to remain at the top of my game to be a safe winch operator. I was supposed to be helping people, not endangering their lives by not doing my job well. After a couple of restless nights' sleep, I came to the most difficult decision of my life. It was time to retire.

I sat down with the CEO and told him that I wanted to step down from being a rescue helicopter crewman at the end of March, around three months away. I wanted to give Life Flight enough time to recruit and train a new crewperson and I also didn't want to finish at 24 years and nine months — I wanted to make it to my twenty-fifth anniversary.

KNOWING THAT I WAS going to be finishing up soon put a whole different light on the job. I enjoyed every flight as much as I did when I was a rookie, enjoying the scenery, the thrill of flying and the thrill of being part of the team. I never knew if we'd get a flight when I was on duty, so I treated every flight like it might be my last.

I continually wondered if I had made the right decision. I knew I was a good crewman and I just needed more flight time. But, knowing how rusty my skills were, it would have been negligent on my part to show up at work and make believe I was still at the top of my game. My core skills were still fine, but accidents happen on the edges. It would have been impossible to look myself in the mirror if I seriously injured or killed one of my teammates, or someone we were trying to rescue, because I let my vanity or sense of self-worth be more important than another person's life.

The twenty-fifth anniversary passed in March but the recruitment process still had not begun. I would check in on a regular basis and was always promised that the recruitment would begin soon and I was asked to keep on flying until the new person was on board.

In April, when I had to call in one of the other crew to winch a sick fisherman off a boat, I took it as a sign that it was time to finish up. I had made the right decision, for the right reasons, and here I was now four months down the track and even less current than when the incident occurred.

I emailed the CEO and told him that I felt I had given them enough time to recruit a new person and, since no one had even been hired, it would still be months before they were trained and ready. I was scheduled to cover a shift on 8 May, and that would be my last one.

I had a lot of support for my decision from within the team. Many of them felt that nothing was progressing with hiring a new crewperson, and if I didn't step down, nothing would. They were sad for me, but saw it as a plus for the operations team.

I HAD MIXED feelings when I arrived at the Air Rescue Centre for my last day. It was incredibly sad to think that this was it, the end of an era. But I took some time to walk around and look at where we were today, compared to when I started 25 years ago. My sadness was tempered by pride in how far Life Flight had come, and how much

Dave in the helicopter, 2014.

End of My Flying

I had helped contribute to its growth over the years.

I ended up having three flights on my final day. It was much better to be busy than to sit around and do nothing.

After my last flight I returned to the hangar and finished my cleaning and paperwork, neither of which I would miss much.

I was thankful that no one else was around when I changed out of my flight suit and put away my helmet for the very last time. The only eyes in the locker room were mine, and there wasn't a dry eye in the house.

Dave's locker after his last flight.

Dave at his 2014 farewell with Steve Creeggan (Anzac Day), Katie-Jane Bowen (motorcycle crash) and Chris Webb (*Terminator*).

Henny Nichols, Dave and Sarah Cody at Dave's farewell.

FORTY-FOUR
CHANGES OVER TIME

I'VE WITNESSED MANY changes over my career; some I liked, some I didn't, some I agreed with and some I was vehemently opposed to. At the end of the day, most change is a good thing and it is going to happen whether we like it or not.

The biggest changes over the past 25 years have been the technology we use and the attitudes towards safety and risk minimisation.

When I started in 1991, the prevailing attitude for missions was 'fill your boots'; that is, do whatever it took to get the job done. By the time I finished up in 2016, the pendulum had swung the other way and we had become so risk aware, that if we couldn't 'check all the boxes', we turned down the mission. Don't get me wrong, I am 100 per cent in favour of doing what needs to be done to keep our teams safe, but, as with all things health and safety related, the truth lies somewhere in the middle.

The early 1990s were still the early days of rescue helicopter services in New Zealand and, without a doubt, they were the most fun. Back then, our pilots were willing to accept most missions and turn back if the weather was unsuitable. I took part in some missions that wouldn't even be considered today. They were adventurous and we accomplished great things and saved a lot of lives, but, with hindsight, I think we were very lucky as an industry that we didn't suffer more accidents.

Early on, the Kiwi No. 8 wire attitude prevailed and some innovative things, like our internal fuelling system, were born. John was not only our Operations Manager but also a very ingenious man who designed and built many of the stretcher systems still used in helicopters and fixed-wing air ambulances today. Back then, he built it, tested it and then we used it. Today, it takes tens of thousands of dollars and many months to get new things installed in the aircraft. Some of the cost and regulations make sense to me, others not so much.

Technological advancements which made the job easier and safer included the night vision goggles and GPS. Today we all take GPS units for granted, but they have only been readily available since 1995. Prior to that we had to be excellent at map and chart reading. Smartphones, tablets and the ability to track aircraft from a computer are examples of technology advances that made a significant difference.

Alongside, medical technology has considerably advanced. Medical gear has become smaller, lighter and more sophisticated, allowing medical teams to monitor patients in the aircraft the same way they could in the hospital.

Our move into the new, custom-designed Air Rescue Centre at the airport was a hugely beneficial change. Now, every time I fly out of Wellington, I am proud to look across and see the Life Flight logo hanging on a building I had so much to do with.

The one thing that has never changed in the rescue services is the passion and dedication of the people involved. And, at the end of the day, it is the people, using the technology, that make the difference.

Change is good and it is necessary, but it is not always fun to go through.

Pilot Mike Hall flies in front of a blue moon.
Andrew Gorrie/Dominion Post

FORTY-FIVE
NEW BEGINNINGS

AS SOON AS I realised that my Operations Manager job was going to come to an end I knew I had to create a new life for myself. I hoped that remaining as a crewman on the helicopter was part of the new life, but there were no guarantees.

As I was figuring out what I wanted to do I spent a lot of time speaking with Lynn, a friend who has been my life coach and counsellor for many years. One of the best and most practical things she had me do was get a few sheets of cardboard and sticky notes. I wrote down my skills, organisations I might want to work with, things I wanted to be involved in, places I might want to live and some short- and long-term goals. Everything that was important to me in my life was represented on these cardboard sheets. These helped me choose what my life looked like, moving forward.

One thing that I was very clear on was that I wanted to remain in Wellington. Wellington is my home; I have more roots here than anywhere else in the world and I had no desire to start a new life somewhere new.

Being of service to others has always been important to me. Everything from my emergency service roles to looking after rugby teams has involved helping others. I knew that I wanted this to continue as I moved forward.

I knew I wanted to stay involved in the SAR and air ambulance sectors if I could; I had years of knowledge and experience that could be put to good use if I found the right outlets.

I was keen to become more involved in emergency management and help people prepare for, survive and thrive before, during and after a natural or man-made disaster. With that in mind, a friend and I set up a company called PrepareMe, to help families and businesses do just that.

I have built up an incredible network of friends and business associates over my years in New Zealand, and I was determined to use this network to move onto this next phase of my life. As I reached out to these people, many opportunities presented themselves.

I met with Duncan and Carl, from the New Zealand Search and Rescue Secretariat (NZSAR), who said they had a project coming up which could be perfect for me. Once my disestablishment was confirmed, I began working with them to organise and facilitate Mass Rescue Operation (MRO) tabletop exercises around New Zealand.

An MRO is a search and rescue incident involving large numbers of people, which overwhelms the capabilities of local authorities. For instance, if a cruise ship with thousands of people on board hit rocks and sank anywhere around New Zealand's coast, as the *Costa Concordia* did in Italy, this would be an overwhelming event.

The tabletop exercises involve emergency services and emergency management agencies, including the Police, RCCNZ, Fire, Ambulance, Civil Defence and many others, to exercise a district's readiness for an MRO. One of the things I have really enjoyed is using 'dynamic simulation' in our tabletop exercises. Dynamic simulation allows the exercise to flow based on what the participants do, not what we, as the organisers, expect them to do. This type of exercise ensures a more realistic outcome for those participating. I now use dynamic simulation in exercises I run for other businesses and government agencies and this has helped bring their incident management exercises to a new level.

The MRO exercises keep me heavily involved with search and rescue and allow me to work with people in the emergency management field. Travelling the country and meeting so many people, during these exercises, has also opened other doors, including spending a week on the Chatham Islands working with their local civil defence group during a National Civil Defence exercise. The exercise was a simulation of a large offshore earthquake, followed by a tsunami, hitting New Zealand. It was great to spend time on the Chathams given that my past visits had literally been flying ones. Ironically, my trip was extended by a day, because the morning I was supposed to fly out a large earthquake struck in the Pacific Ocean and many parts of the country, including the Chathams, were under a real tsunami threat. Luckily, a large tsunami didn't eventuate!

I have also been tutoring air observation techniques to helicopter and fixed-wing crews, as well as first-aid and CIMS, New Zealand's incident management system, to agencies around the country.

Given my dislike of school in the past, I surprised myself when I decided to go back to school and obtain a graduate degree in emergency management. I am currently taking one class each semester and am on my way to completing the degree. To my amazement, I have managed to achieve an A or B+ grade for all my classes so far. I am enjoying it so much that I sometimes think about completing a master's degree!

I have also learned a big lesson about how and where I like to work. I set up an office in my apartment and thought it was working well. Soon I realised that I was spending a lot of time walking to and from Prefab, my favourite café, and not getting nearly as much work done as I thought I would.

I started basing myself in a co-working space called Bizdojo which has helped me bring my business to a new level. Instead of working in the solitude of my apartment, I go to work each day surrounded by 200 creative, hardworking people in an amazing workspace. We all get on doing our own things, but I always have someone nearby to give me a second opinion, or to share a win with. I think that finding a co-working space is one of the best things that has happened to improve my business.

ALL THE SELF-HELP books tell you to find a job you love and you will never work another day in your life. I haven't seen any books on what to do when your dream job gets ripped out from under you! After having lived my passion for so long, one of the biggest challenges I face now is staying motivated and finding new things to be passionate about.

I avoid self-pity because I know that a lot of people face this type of dilemma every day. Every pilot is one medical away from losing their licence, and every sportsperson is one injury away from forced retirement. Knowing that other people face the same dilemma is not comforting, it just makes me feel sad for them, too!

I'd like to find ways to help other people going through a similar experience, so they come out stronger on the other side.

I LIVE NEAR Wellington Hospital so I often hear the helicopter flying, and wonder who is on board and what they have been up to. Every single day I miss flying and I often ponder if I should have fought harder to keep on doing it. After a few minutes of quiet consideration, I realise I did make the right decision. I wanted to be there, but the powers-that-be didn't want me.

I am often asked if the management changes that the Life Flight Trustees made in 2014 were the right thing to do, and my honest response is that I have no idea. However, I do know that the operational teams are still out there saving lives every single day, and that is all that counts!

I hope that one day I am invited to join the board of The Life Flight Trust. My knowledge and passion for Life Flight, and the air ambulance/air rescue industry remains strong, and I think I could be a real asset for them.

UNFORTUNATELY, I FEEL like my emergency service days are over.

A real sense of helplessness and frustration came upon me when the Kaikoura earthquake struck in November 2016. I was lying in bed as the longest earthquake I have experienced shook my apartment. I laid there, wondering if this was the 'big one', the earthquake that would bring Wellington to its knees.

When the shaking stopped I jumped out of bed, ready to go and assist someone, anyone. Suddenly I felt terribly frustrated that, for the first time in 25 years, I didn't have an official role to play in any response to the emergency. I was ashamed for having such a selfish reaction to the event, but it is all part of getting used to a life outside the emergency services.

MOVING FORWARD A BIG dream is to use my 42 years of experience in the emergency services, and my love of being 'in front of the room' to embark on a new career of motivational or after-dinner speaking. I am an amusing and informative speaker who can entertain a crowd with a good yarn, or use my helicopter missions as a metaphor for life and business.

I have found the trick is to tell a good, memorable story, and then help a person or team see how they can apply those lessons in their own world. I have hundreds of real-life examples of teamwork, resilience, working under stress, making decisions without all the required information to hand and many other subjects.

Crew Resource Management is so important to air crews and I believe I can get out there and add a new perspective to how it is taught. Most of the CRM courses I have attended were taught by a pilot, from the pilot's point of view. I think there are important lessons on also seeing CRM from a non-pilot point of view. If you are a helicopter paramedic trying to save a patient's life, how do you balance the safety of the patient with the safety of the crew?

Another of my goals is to find a way to work with and help Stevin Creeggan, the survivor of the Anzac Day crash. I have never met a man with more resilience and courage and I think we would make an awesome presenting team.

I ALSO FEEL like I have more writing to do! If you've liked what you have read in this book then sign up to my mailing list at: http://www.rescuedave.com/emergency-response

Everyone who signs up will be sent the full story of how winching an injured tree surgeon, from the top of a tree, didn't go as well as it might have!

OVER THE YEARS, I have seen enough bad things happen to good people that I know life is precious and it is also unpredictable. I strongly believe that we get to shape our own destinies, and no matter what happens in life, it is our reaction to it, not the event itself, which defines us.

So, for me, when I'm not out presenting, tutoring or consulting, I intend having many more adventures in my life, including a lot more fishing, exercising, travelling and hanging out with my friends and family.

Life is short, I intend to live it to its fullest.

DPG

Would you like to see more photos, videos and media articles related to stories in the book? Sign up to my mailing list at http://www.rescuedave.com/emergency-response

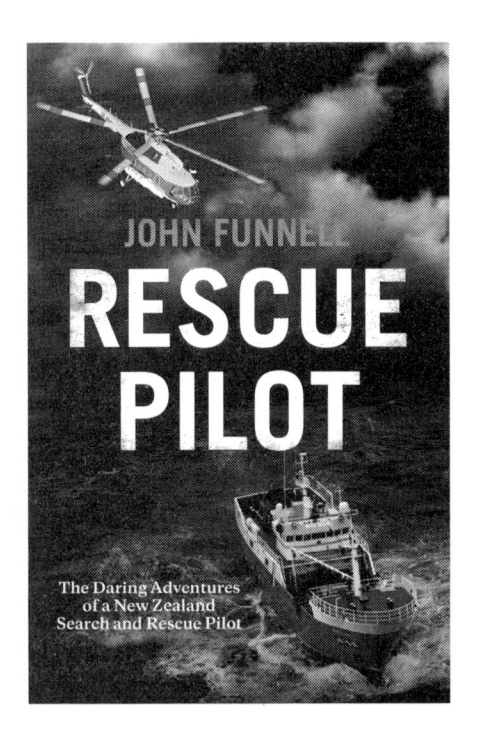

Often referred to as a 'search and rescue daredevil', John Funnell is one of New Zealand's longest serving and most respected search and rescue pilots, having clocked an incredible 19,000 hours of flying time.

In *Rescue Pilot* he shares stories from over four decades of flying search and rescue helicopters all over New Zealand — and beyond. John is a hero to the thousands of victims he has transported to safety over the years, and is perhaps best known for his unprecedented 800-kilometre mission to save a MetService employee attacked by a shark on the remote subantarctic Campbell Island.

Fearless and prepared to give pretty much any mission a go, John boasts a remarkable aviation career. What's more, he's a natural-born story-teller, and the adventures he recounts are utterly gripping.

For more information about our titles please visit
www.penguin.co.nz